SpringerBriefs in Applied Sciences and Technology

Manufacturing and Surface Engineering

Series Editor

J. Paulo Davim

Bekir Sami Yilbas

Laser Drilling

Practical Applications

 Springer

Bekir Sami Yilbas
Mechanical Engineering Department
King Fahd University of Petroleum
    and Minerals
Dhahran
Saudi Arabia

ISSN 2191-530X   ISSN 2191-5318   (electronic)
ISBN 978-3-642-34981-2   ISBN 978-3-642-34982-9   (eBook)
DOI 10.1007/978-3-642-34982-9
Springer Heidelberg New York Dordrecht London

Library of Congress Control Number: 2012953192

Printed on acid-free paper

Springer is part of Springer Science+Business Media (www.springer.com)

*To my mother Ayse, my wife Zahide,
and my daughters Ayse and Merve*

Preface

Machining is one of the important areas in the engineering discipline. To meet today's challenges, it is necessary to incorporate advanced machine tools in manufacturing processes. Laser drilling is considered to be one of the advanced machining processes filling the gap in the advanced manufacturing systems because of their precision, low cost, localized processing, and high speed of operation. In laser drilling applications, a laser beam is used as a heat source increasing temperature rapidly to the melting and evaporation temperature of the substrate material. Since the arrangements of the optical setting for the laser beam are very precise, the localized heating can be controlled easily. With recent advancement in laser technology and computations power, laser drilling application has become almost an integral part of the aerospace, power, electronic, and sheet metal forming industries. In laser drilling operations, the physical processes are complicated in nature and they require a deep understanding of the process to secure improved end-product quality.

In laser drilling processes, the end-product quality is very important from the manufacturing point of view. The quality assessment of the end product, such as drilled hole, can be possible through examining the geometric features of the laser drilled hole section. One of the methods associated with the quality assessment is the factorial analysis; in which case, the affecting factors are varied randomly or with increments to generate the random blocks. Through the statistical testing of the measurable responses, the significant levels of the affecting parameters can be identified.

In this book, laser drilling operation is introduced and analysis related to the drilling mechanisms is presented in Chap. 2. The qualitative and quantitative analysis of laser drilled holes is accomplished in Chap. 3 to assess the end-product quality. Chapter 4 deals with the above surface phenomena, which influence the end-product quality. In the last chapter concluding remarks are included for the laser drilling process.

Acknowledgment

I would like to thank my family for their courage and kind support to make this book possible. I would like acknowledge the role of King Fahd University of Petroleum and Minerals in extending strong support from the beginning to the end facilitating every means during the preparation of the book. The author wishes to thank the colleagues who contributed to the work presented in the book through previous cooperation of the author. In particular, thanks to Dr. Muammer Kalyon, Dr. Mehmet Pakdemirli, Dr. Hussain Al-Qahtani, Dr. Nasser Al-Aqeeli, Dr. Shahzada Zaman Shuja, Dr. Saad Bin Mansoor, Dr. Ahmad Al-Dweik, Dr. Abdul Aleem B. J., and to all my graduate students.

Contents

Chapter 1
Introduction

Lasers can be considered non-conventional machine tools and find applications in industry because of their precision of operation, low cost, localized processing, and high speed of operation. In laser drilling applications, a focused laser beam is used as heat source increasing temperature rapidly to the melting and evaporation temperature of the substrate material. Since the arrangements of the optical setting for the laser beam is very precise, the localized heating can be controlled easily and desired hole features can be accomplished with maximum accuracy and minimum defects. Laser drilling can be categorized into non-conduction limited heating, since the phase change processes, including melting and evaporation, are involved during the drilling. Although the laser drilling appears to be the simplest laser machining operation, the drilling involves with the complicated physical processes. This is due to phase changes occuring at high temperatures and at short time intervals. In order to optimize the laser machining process and reduce the experimental time and cost, the model studies receive considerable attention. In addition, the model studies give insight into the physical processes that take place during the heating process and being easier to accomplish as compared to experimental studies. The measurement of physical properties during laser workpiece interaction is difficult and costly since, the process is involved with high temperature, short duration, and localized heating. From the modeling point of view, laser drilling can be fallen into two-dimensional axis symmetric non-conduction limited heating category. Analytical solution to the drilling problem becomes complicated and difficult to achieve. Therefore, numerical treatment of the heating problem is easier and fruitful. When the high power laser beam focused onto the substrate surface, the beam energy is partially absorbed by the substrate material. Depending on the focused beam diameter at the surface, laser power intensity (combining the laser output energy and pulse length), and reflectivity of the surface, substrate material undergoes solid heating, melting and evaporation. In the case of evaporation process, the evaporating front detaches from the liquid surface generating a recoil pressure across the vapor–liquid interface. As the evaporation of the surface progresses, the recoil pressure increases considerably while influencing

B. S. Yilbas, *Laser Drilling*, SpringerBriefs in Manufacturing and Surface Engineering, DOI: 10.1007/978-3-642-34982-9_1, © The Author(s) 2013

the evaporation rate. As the heating progresses further, the liquid surface recesses towards the solid bulk forming the cavity in the substrate material. Depending on the pulse length and power intensities, the liquid ejection from the cavity occurs, which is particularly true for the long pulses (~ms pulse lengths); however, the surface ablation without liquid ejection takes place for short pulses (~ ns pulse lengths). The liquid ejection improves the material removal rates from the cavity. In the case of laser short pulse processing, the recoil pressure increases substantially due to high rates of momentum exchange during the evaporation process. In this case, high pressure at the vapor/liquid interface acts as a pressure force generating a surface stresses at the liquid/solid interface. This, in turn, results in a pressure wave propagating into the substrate material. Depending on the magnitude of pressure wave, the plastic deformation through dislocations in the surface region of the substrate material takes place. The depth of deformed region is limited with the interaction of loading (plastic wave) and unloading (elastic wave) waves., i.e. as the loading phase is completed (when the evaporation is completed, the recoil pressure diminishes), unloading wave (elastic wave) from the liquid/solid interface initiates. Since the unloading wave travels faster than the loading wave, both waves meet at some depth below the surface. It should be noted that the wave motion in the substrate material is complicated and requires comprehensive investigation.

In laser drilling process, the end-product quality is very important from the manufacturing point of view. Consequently, the optimization studies improve laser drilling process considerably. The quality assessment of the end product, such as drilled hole, can be possible through examining the geometric features of the laser drilled section. Laser drilled hole quality can be judged by internal form and taper and related geometrical features, as well as extend of heat affected zone. Consequently, the levels of the drilling parameters resulting in holes with less taper and parallel-sided walls are needed to be identified. The laser and workpiece material parameters affecting the laser hole drilling include laser output energy, pulse length, focus setting of focusing lens, drilling ambient pressure and workpiece thickness. The optimum performance in laser drilling depends on the proper selection of these factors. However, with many variables and incomplete information of the relationship between them, a statistical method is fruitful to design an experiment and analyze the results accordingly. One of the methods associated with the quality assessment is the factorial analysis. In this case, the affecting factors are varied at randomly or with increments to generate the random blocks. Through the statistical testing of the measurable responses, the significant levels of the affecting parameters can be identified.

Considerable research studies were carried out to investigate the laser hole drilling process. When modeling the laser drilling, the main emphasis was given to the laser heating and the material ejection processes during the laser-workpiece interaction. The heating mechanisms in relation to the drilling were examined extensively in the past. Some of these studies include the modeling of material response to the laser pulse [1–3]. Moreover, a number of investigators pointed out that typically 90 % of material ejected from the workpiece was in liquid state when the laser power intensity was on the order of 10^{11} W/m^2 [4–6]. Previously, information on the ejected material was recorded photographically of material spatter on slides placed between the irradiated target and the laser beam focusing lens [7, 8]. The measurements provided quantitative description of the material removal during the drilling process; however, it lacks the qualitative assessment of the drilling. Moreover, the drilling process is, in general, complicated in nature and requires further investigations.

Laser drilling process was investigated to reduce the drilling time, minimizing the energy loss, and optimize the end product quality. The effects of focus setting on resulting mean hole diameter were studied by Yilbas [9]. He showed that the increasing laser energy and moving the focus position slightly above the workpiece surface increased the mean hole diameter. Laser drilling into electronic components was studied by Taneko et al. [10]. They showed that use of high peak pulsed CO_2 laser improved the drilling quality. Laser drilling of microvias in epoxy-glass printed circuit boards was investigated by Kestenbaum et al. [11]. They indicated that holes of 0.1 mm diameter could be drilled with a drilling rate of several hundreds per second. Yilbas [12] introduced a parametric study, based on the statistical approach to analyze the most affecting parameters in laser hole drilling process. He mentioned that main effects of laser parameters as well as their interactions had significant effects on features of hole geometry. A study into the effect of beam waist position on hole formation in laser drilling process was carried out by Yilbas and Sami [13]. They showed that altering the beam waist position tends to produce converging and diverging hole walls for negative and positive positions, respectively. The drilling improved at a certain beam waist position for a given workpiece thickness. In addition, Yilbas [14] investigated the laser hole drilling process for various engineering metals. He indicated that the parameter, which was found to be very significant in most cases was the workpiece thickness while first order interactions of pulse length-thickness was the most significant, and pulse length-focus setting was significant for all the materials examined. Kenny and Dally [15] studied laser drilling of very fine electronic via holes in common circuit board materials. They showed that energy delivered was the most critical parameter and indicated that further investigation in hole quality was necessary.

References

1. Duley WW (1983) Laser processing and analysis of materials, 1st edn. Plenum Press, New York
2. Hoadley AFA, Rappaz M, Zimmermann M (1991) Heat-flow simulation of laser remelting with experimental validation. Metall Trans B 22B:101–109
3. Yilbas BS, Sahin AZ, Davies R (1995) Laser heating mechanism including evaporation process initiating the laser drilling. Int J Mach Tools Manufact 35(70):1047–1062
4. Wei SP, Chian LR (1988) Molten metal flow around the base of cavity during a high energy beam penetrating process. Int J Heat Mass Transf 110:918–923
5. Yilbas BS (1998) Particle ejection during laser drilling of engineering metals. Lasers Eng 7:57–67
6. Yilbas BS (1995) Study of liquid and vapor ejection processes during laser drilling of metals. J Laser Appl 17:147–152

7. Yilbas BS, Sami M (1997) Liquid ejection and possible nucleate boiling mechanisms in relation to laser drilling process, J. Phys Part D, J Appl Phys 30:1996–2005
8. Yilbas BS (1985) The study of laser produced plasma behavior using streak photography. Jpn J Appl Phys 24:1417–1420
9. Yilbas BS (1988) The examination of optimal focus setting with material thickness at atmospheric pressures of air and oxygen in laser drilling of metals, Proc Instn Mech Engrs Part B, J Eng Manuf 202:123–127
10. Taneko S, Massahura M, Seigo H (1990) Laser drilling by peak pulsed CO_2 laser. ICALEO' 90 Laser Material Processing, Orlando
11. Kestenbaum A, D'Amico JF, Bulmenstock BJ, DeAngelo MA (1990) Laser drilling of microvias in epoxy-glass printed circuit boards. IEEE Trans Componen Hybrids Manuf Technol 13:1055–1062
12. Yilbas BS (1997) Parametric study to improve laser hole drilling process. J Mater Process Technol 70(1–3):264–273
13. Yilbas BS, Sami M (1996) Study into the effect of beam waist position on hole formation in the laser drilling process, Proc Instn Mech Engrs Part B, J Eng Manuf 210:271–277
14. Yilbas BS (1987) Study of affecting parameters in laser hole drilling of sheet metals. ASME J Eng Mater Technol 109:283–287
15. Kenney AL, Dally JW (1998) Laser drilling of very small electronic via holes in common circuit board materials. Circuit World 14:31–36

Chapter 2
Thermal Analysis of Laser Drilling Process

In laser drilling process, the material removal involves with evaporation at the surface, liquid ejection, and solid heating. However, liquid ejection vanishes for the lasers with the pulse length within the rage and higher than the nanoseconds. However, in drilling applications mass removal by liquid ejection is desirable because the rate of material removed becomes high. Analytical modeling the laser heating process in relation to drilling is difficult, since process involves with the phase change and fluid flow due to the evaporation at the surface. However, numerical modeling is feasible with some useful assumptions.

2.1 Heating Analysis

Laser high intensity beam interaction with the solid surface results in rapid evaporation of the surface. Depending on the power intensity and the duration of the laser pulse, the pressure generated at the laser irradiated surface becomes very high. This situation is especially true for nanosecond laser pulses. Moreover, the prediction of recoil pressure is essential, since the mass removal rate depends on the pressure differential at the cavity surface. Consequently, one of the governing physical parameters involving laser ablation is the pressure generated at the interface of vapor–liquid phases in the cavity. Since the pulse duration is very short and the magnitude of pressure generated in a small area (limited to irradiated spot size) is very high, liquid ejection replaces the vapor jet emanating from the irradiated surface. Since the process is rapid and involves with high temperature phenomenon experimentation into the physical processes becomes difficult and expensive. However, model studies give insight into the physical processes taking place during the ablation process. The analyses related to the model study are presented in line with the previous studies [1–5].

To model the laser evaporative heating situation, energy equation for each phase needs to be solved independently as well as coupled across the interfaces of the two-phases mutually exist (mushy zones). In the initial stage of heating,

B. S. Yilbas, *Laser Drilling*, SpringerBriefs in Manufacturing and Surface Engineering, DOI: 10.1007/978-3-642-34982-9_2, © The Author(s) 2013

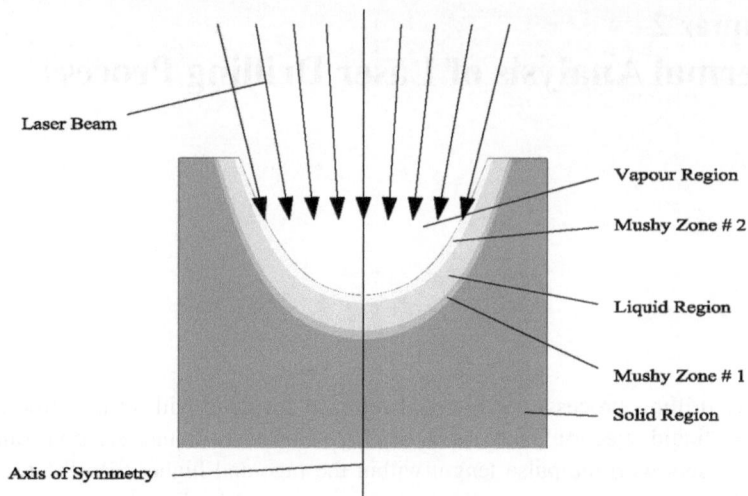

Fig. 2.1 Schematic of a laser drilling process

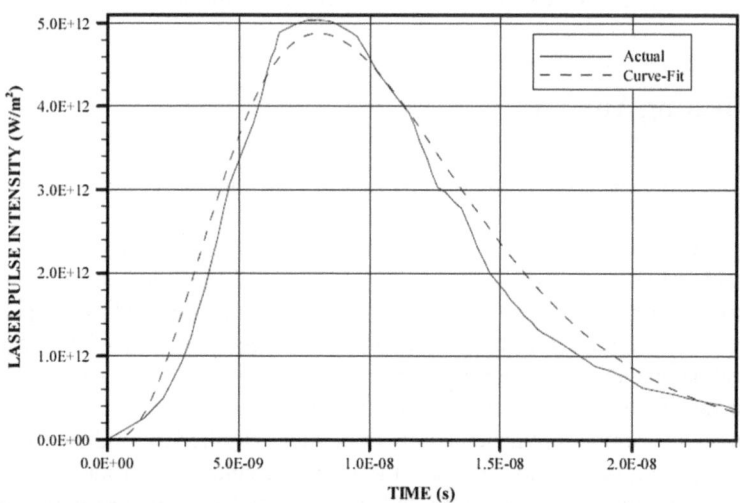

Fig. 2.2 Temporal variation of laser pulse intensity, obtained from equation and the measured from the experiment

conduction in solid with convective boundary at the surface should be considered. In actual laser pulse heating, the laser output power intensity distribution at the surface of the workpiece is Gaussian and this should be accommodated in the analysis such that its centre is at the centre of the co-ordinate system Fig. 2.1. The temporal variation of laser power intensity resembling the actual laser pulse is shown in Fig. 2.2. This arrangement results in an axisymmetric heating of the substrate material. The diffusion equation for a solid phase heating due to a laser irradiation pulse with a Gaussian intensity profile can be written as:

2.1.1 Solid Phase Heating and Phase Change

Let us first consider the solution procedure of a 2D melting problem. We start from a solid body whose temperature is rising with time in a nonuniform manner. Energy is provided by a nonuniform volumetric heat source inside the body, which causes the temperature to rise in the body. Due to the nonuniform nature of the heat source the temperature distribution inside the body becomes also nonuniform. As heating progresses, a time is reached when the temperature at some parts of the body reaches melting temperature of the substrate material and the material converts into a liquid phase. One can tackle this problem in the following manner. Start solving the unsteady heat conduction equation in the solid phase. In the time step, in which some parts of the body reach or exceed the melting point, the boundary between solid and potential liquid regions should be determined. Since a fixed rectangular grid is chosen in the solution domain, there are regions where the solid–liquid interface will pass through the individual cells. In this case, all such cells make up a so-called 'mushy zone' in which they are in partly solid and partly liquid phases. In this scenario, the usual heat conduction equation cannot be solved in these regions, but instead a 'quality' equation should be considered.

The quality of a mushy zone (cell) can range from 0 to 1. A quality of 0 designates a solid cell and a quality of 1 designates a liquid cell. Therefore, in the newly demarcated region calculate the quality of the cells by solving the quality equation. There will be cells whose quality will be in between 0 and 1. All such cells make up the 'mushy zone'. There will be cells whose quality will be equal to or greater than 1. All such cells make up the liquid region. Calculate the boundaries of these two regions. In the liquid region one should solve the heat conduction equation to obtain the temperature field. This procedure is repeated in the next time step. In this way one can proceed in time to determine the region where phase change takes place. As simulation progresses the time step will be reached in which the temperature of some parts of the liquid region will reach or exceed the boiling point of the substrate material. At this point, we start calculating the boundary of the liquid and the potential vapor regions. In the potential vapor region we can then calculate the quality of each cell to determine whether it has fully converted to vapor phase or not. There will be cells whose quality will be in between 0 and 1. All such cells make up the second 'mushy zone' region. There will be cells whose quality will be equal to or greater than 1. All such cells make up the vapor region. One should calculate the boundaries of these two regions. The boundary between the second mushy zone and the vapor region actually defines the cavity shape. This procedure is repeated for all future time steps. In this way the cavity shape can be predicted.

For modelling a laser drilling operation, one approach is to assume the laser heating by means of a volumetric heat source, which is spatially and temporally distributed in the workpiece. The magnitude of the volumetric heat source depends on the laser surface intensity, the laser beam absorption depth and the reflectance of the laser beam from the surface. Let us now list the major assumptions of the mathematical model employed in the analysis.

- Fourier's law of heat conduction is applicable in the solid and liquid regions.
- The problem is 2D axisymmetric.
- The material is a pure substance with single melting and evaporation temperatures.
- Laser absorption is modelled by a volumetric heat source in the metal.
- Thermo-physical properties of the metal are constant.
- The vapor does not interact with the laser beam.
- Liquid movement and expulsion are negligible.
- Insulated boundary conditions at the upper surface.
- Solid and liquid phases have the same absorption coefficient.
- Material has a certain absorption depth for the laser radiation.
- Laser beam intensity has Gaussian distribution.
- There is no ionization of the emerging gas front.
- There is no multiple reflection phenomena from the surface.
- Laser pulse is time exponentially varying resembling the actual laser pulse intensity.

Mathematical details of the problem are presented in the next section. We have three distinct stages. In stage 1 the temperature anywhere in the solid region is below melting point. In stage 2 the temperature anywhere in the liquid is below evaporation temperature. In stage 3 we have three distinct phases, solid, liquid and vapor.

The transient heat transfer equation for a solid substrate at constant properties irradiated by a laser beam is presented below.

Solid Phase Heating:

In the solid phase, the Fourier heat conduction equation is used. This is due to that the length and the time scales considered in the analysis are larger than interatomic spacing and electron relaxation time. Therefore, the heat diffusion equation is:

$$\rho_s C p_s \frac{\partial T_s}{\partial t} = k_s \left[\frac{1}{r} \frac{\partial}{\partial r} \left(r \frac{\partial T_s}{\partial r} \right) + \frac{\partial^2 T_s}{\partial z^2} \right] + S \tag{2.1}$$

where T_s is the temperature in the solid phase and the time dependent laser source term S is:

$$S = I_o (t) \delta \left(1 - r_f \right) e^{(-\delta z)} e^{-(r/a)^2} \tag{2.2}$$

However, peak power intensity at the workpiece surface can be formulated to resemble the actual laser pulse. In this case, the following equation is adopted for peak power intensity.

$$I_o = \left[cp_1\left(t/cp_5\right) + cp_2\left(t/cp_5\right)^2 + cp_3\left(t/cp_5\right)^3\right]\exp\left(cp_4\left(t/cp_5\right)\right) \quad (2.3)$$

It should be noted that the coefficients cp_1, cp_2, cp_3, cp_4, and cp_5 can be obtained from the actual pulse. For example, for the laser pulse shape shown in Fig. 2.2, these coefficients are:

$$
\begin{aligned}
I_o &= 10^{13} \ (Wm^{-2}) & cp_1 &= 0.0105771617002674 \\
\delta &= 6.17 \times 10^6 \ (m^{-1}) & cp_2 &= 0.0118181307770856 \\
r_f &= 0.5 & cp_3 &= 750.004377930276 \\
a &= \tfrac{2r_o}{3} \ (m) & cp_4 &= 10.1106779350279 \\
r_o &= 1.25 \times 10^{-5} \ (m) & cp_5 &= 2.71e - 008
\end{aligned}
$$

Initial Condition:
Initially material is considered at uniform ambient temperature, which is 300 K.

$$T_s\left(z,r,t\right) = 300 \ K \ at \ t = 0$$

Boundary Conditions:
At a depth of infinity, temperature is assumed to be reduced to initial temperature (300 K).

$$T_s\left(z,r_{\max},t\right) = 300 \ K$$

$$T_s\left(z_{\max},r,t\right) = 300 \ K$$

At symmetry axis, temperature is assumed to be the maximum due to axis-symmetry heating situation.

$$\left.\frac{\partial T_s}{\partial r}\right|_{\substack{z=z \\ r=0 \\ t=t}} = 0$$

At the surface, convective boundary is assumed with $h \cong 10^2 \ W / \left(m^2 K\right)$ and T_o is the temperature at infinity, which is 300 K.

$$\left.-k_s\frac{\partial T_s}{\partial z}\right|_{\substack{z=0 \\ r=r \\ t=t}} = h\left(T_{surface} - T_o\right)$$

where, I_o, δ, r_f and a are the laser power intensity, reciprocal of the absorption depth, reflectance and the Gaussian parameter, respectively. cp_1, cp_2, cp_3, cp_4 and cp_5 are curve fit constants. The numerical values of the different parameters are:

In the time step in which the temperature, in some cells, reach or exceed the melting point we have to start calculating the phase change boundaries as well as the quality in the mushy zone. In the cells where the quality, x_m reaches or exceeds 1 an unsteady heat conduction equation is also to be solved for the concerned cells. The quality equation is derived by means of the energy method. Consider a differential element in a substrate material, which is subjected to a melting process and let x_m be the mass fraction of the liquid present in the element, then the energy content (ΔU) of the differential element with volume (ΔV) at the melting temperature T_m can be written as:

$$\Delta U = \rho_m \Delta V \left[x_m \left(L_m + Cp_m \left(T_m - T_{ref} \right) \right) + Cp_s \left(1 - x_m \right) \left(T_m - T_{ref} \right) \right] \quad (2.4)$$

where

$$x_m = \frac{m_m}{m_m + m_s} \quad (2.5)$$

and, T_{ref}, x_m, m_m and m_s are the reference temperature for enthalpy, quality of liquid, mass of liquid and mass of solid in the element, respectively. After assuming that specific heat of melt is the same as solid at the melting temperature ($Cp_s = Cp_m$ at $T = T_m$), the above equation reduces to:

$$\Delta U = \rho_m \Delta V \left[x_m L_m + Cp_m \left(T_m - T_{ref} \right) \right] \quad (2.6)$$

For a unit volume, it reduces to:

$$\frac{\Delta U}{\Delta V} = \Delta u = \rho_m \left[x_m L_m + Cp_m \left(T_m - T_{ref} \right) \right] \quad (2.7)$$

Differentiation with time yields:

$$\frac{\partial u}{\partial t} = \rho_m L_m \frac{\partial x_m}{\partial t} \quad (2.8)$$

since $Cp_m \left(T_m - T_{ref} \right) = $ const.

It is important to note that in conduction Eq. 2.1, $\rho_s Cp_s T_s$ is also the enthalpy per unit volume i.e.

$$\rho_s Cp_s \frac{\partial T}{\partial t} = \frac{\partial u}{\partial t} \quad (2.9)$$

Substituting Eq. (2.9) into (2.1) gives the energy equation for the differential element subjected to the phase change process (melting):

$$\rho_m L_m \frac{\partial x_m}{\partial t} = k_m \left[\frac{1}{r} \frac{\partial}{\partial r} \left(r \frac{\partial T}{\partial r} \right) + \frac{\partial^2 T}{\partial z^2} \right] + S \quad (2.10)$$

Equation (2.10) is applicable for the differential elements (cells defined by nodes in the substrate material) when temperature becomes melting temperature of the substrate material $T = T_m$ and $0 \le x_m \le 1$, i.e., a mushy zone. Consequently, here temperature of the cells with $0 \le x_m \le 1$ is set to melting temperature ($T = T_m$). When the value x_m exceeds 1, ($x_m > 1$) and Eq. (2.10) is not applicable for the differential element under consideration. In this case, Eq. (2.1) is used to determine the temperature rise in the liquid heating with the liquid thermal properties employed, i.e., the liquid heating initiates and continues till the temperature reaches the evaporation temperature. It is important to note that inside the mushy zone terms like $\frac{\partial T}{\partial r}$ and $\frac{\partial T}{\partial z}$ are zero, because temperature is constant, but Eq. (2.10) is valid at mushy zone/solid and mushy zone/liquid interfaces where these terms are not generally zero.

Let us now discuss the time step in which the temperature in some cells reach or exceed the boiling point. In this case again a mushy zone arises whose constituent cells are part vapor and part liquid. Equation (2.10) is valid in this second mushy zone when the appropriate thermophysical properties are used in it. The appropriate equation is:

$$\rho_b L_b \frac{\partial x_b}{\partial t} = k_b \left[\frac{1}{r} \frac{\partial}{\partial r} \left(r \frac{\partial T}{\partial r} \right) + \frac{\partial^2 T}{\partial z^2} \right] + S \qquad (2.11)$$

Equation (2.11) is applicable for the range $T = T_b$ and $0 \le x_b \le 1$ in the mushy zone (partially liquid and partially vapor). Consequently, temperature of the cells with $0 \le x_b \le 1$ is set to boiling temperature $(T = T_b)$. It should be noted that x_m is replaced with x_b, which represents the fraction of vapor phase in the differential element.

Melting without Evaporation:

In the second stage three distinct regions exist; solid, solid–liquid mushy zone and liquid. Three different differential equations are to be solved, one in each phase. In the solid and the liquid phases the unsteady heat conduction with heat generation is to be solved, each equation incorporating the appropriate thermophysical properties. In the solid–liquid mushy zone Eq. (2.10) is to be solved. It is to be noted that these regions are not fixed in space but move with time. So that in each time step, before solving Eqs. (2.10) and (2.11), the boundaries of these regions have to be calculated according to the following criterion.

$$T_s \ge T_m \rightarrow Solid - liquid \, mushy \, zone$$
$$x_m \ge 1 \rightarrow Liquid \, region$$

In addition, it should be mentioned that within this stage 2 there would be a time duration in which only the solid and solid–liquid mushy zone will exist. The relevant equations along with the boundary conditions are listed below:

Solid Phase:

Since the length and time scales are larger than the interatomic spacing, therefore the heat diffusion Eq. 2.1 is used with the boundary conditions, except at the solid and solid–liquid mushy zone boundary, the temperature is understood to be the melting temperature.

$$T_s (z, r, t) = T_m \, at \, solid \, and \, solid - liquid \, mushy \, zone \, interface$$

Solid–Liquid Mushy Zone:

In the solid–liquid mushy zone the quality (x_m) is calculated by means of the following equation.

$$\rho_m L_m \frac{\partial x_m}{\partial t} = k_m \left[\frac{1}{r} \frac{\partial}{\partial r} \left(r \frac{\partial T}{\partial r} \right) + \frac{\partial^2 T}{\partial z^2} \right] + S \qquad (2.12)$$

Initial Condition:

Initially the substrate material is all solid, therefore the quality x_m at every node is 0.

$$x_m\,(z,r,t) = 0 \, at \, t = t_m$$

where t_m is the time at which melting starts in the solid substrate.

Boundary Conditions:

At symmetry axis, the quality x_m is assumed to be the maximum due to axisymmetric heating situation.

$$\left.\frac{\partial x_m}{\partial r}\right|_{\substack{z=z \\ r=0 \\ t=t}} = 0$$

At the surface the gradient in the z-direction is assumed to be zero.

$$\left.\frac{\partial x_m}{\partial z}\right|_{\substack{z=0 \\ r=r \\ t=t}} = 0$$

At the solid and solid–liquid mushy zone boundary, the quality is 0.

$$x_m\,(z,r,t) = 0 \, at \, solid \, and \, solid - liquid \, mushy \, zone \, interface$$

At the liquid and solid–liquid mushy zone boundary, the quality is 1.

$$x_m\,(z,r,t) = 1 \, at \, liquid \, and \, solid - liquid \, mushy \, zone \, interface$$

Liquid Phase:

Since the length and time scales are larger than the interatomic spacing, therefore the heat diffusion equation is used:

$$\rho_l C p_l \frac{\partial T_l}{\partial t} = k_l \left[\frac{1}{r}\frac{\partial}{\partial r}\left(r\frac{\partial T_l}{\partial r}\right) + \frac{\partial^2 T_l}{\partial z^2} \right] + S \qquad (2.13)$$

where T_l is the temperature in the liquid phase.

Initial Condition:

Initially the liquid phase is at a uniform temperature, which is the melting temperature $T_m = 1811\,K$. Therefore, the initial condition is:

$$T_l\,(z,r,t) = T_m \, at \, t = t_{sl}$$

where t_{sl} is the time at which the solid–liquid mushy zone starts converting into the liquid phase.

Boundary Conditions:

At symmetry axis, the temperature T_l is assumed to be the maximum due to axisymmetric heating situation.

$$\left.\frac{\partial T_l}{\partial r}\right|_{\substack{z=z \\ r=0 \\ t=t}} = 0$$

At the surface, convective boundary is assumed with $h \cong 10^2 \, W/\left(m^2 K\right)$ and T_o is the temperature at infinity, which is 300 K.

$$-k_l \left.\frac{\partial T_l}{\partial z}\right|_{\substack{z=0 \\ r=r \\ t=t}} = h\left(T_{surface} - T_o\right)$$

At the liquid and solid–liquid mushy zone boundary the temperature is taken to be the melting temperature.

$$T_l\left(z,r,t\right) = T_m \, at \, liquid \, and \, solid - liquid \, mushy \, zone \, interface$$

Initiation of Evaporation:

In the third stage all three phases exist as well as two mushy zones. Five differential equations are to be solved, one in each region along with the appropriate initial and boundary conditions. Again it should be stated that liquid–vapor mushy zone and the vapor region move with time so that before calculating the quality and temperature in these regions respectively, their boundaries should be calculated according to the following criterion.

$$T_l \geq T_b \rightarrow Liquid - vapor \, mushy \, zone$$
$$x_b \geq 1 \rightarrow Vapor \, region$$

As in the second stage, the third stage also consists of time duration in which only the solid, solid–liquid mushy zone, liquid and liquid–vapor mushy zone regions exist and there is no vapor region. The relevant equations along with the boundary conditions are listed below.

Liquid Phase:

In the liquid phase, Eq. (2.13) is used with appropriate boundary conditions, except at the liquid and solid–liquid mushy zone boundary the temperature is taken to be the melting temperature.

$$T_l\left(z,r,t\right) = T_m \, at \, liquid \, and \, solid - liquid \, mushy \, zone \, interface$$

At the liquid and liquid–vapor mushy zone boundary the temperature is taken to be the evaporation temperature.

$$T_l\left(z,r,t\right) = T_b \, at \, liquid \, and \, liquid - vapour \, mushy \, zone \, interface$$

Liquid–Vapor Mushy Zone:

In the liquid–vapor mushy zone the quality (x_b) is calculated by means of the following equation.

$$\rho_b L_b \frac{\partial x_b}{\partial t} = k_b \left[\frac{1}{r}\frac{\partial}{\partial r}\left(r\frac{\partial T}{\partial r}\right) + \frac{\partial^2 T}{\partial z^2}\right] + S \qquad (2.14)$$

Initial Condition:

Initially the cells in the vapor–liquid mushy zone are all liquid, therefore the quality x_b at those nodes is 0.

$$x_b = 0 \, at \, t = t_b$$

where t_b is the time at which evaporation starts in the liquid region.

Boundary Conditions:

At symmetry axis, the quality x_m is assumed to be maximum due to axi-symmetric heating situation.

$$\frac{\partial x_b}{\partial r}\bigg|_{\substack{z=z \\ r=0 \\ t=t}} = 0$$

At the surface the gradient in the z-direction is assumed to be zero.

$$\frac{\partial x_b}{\partial z}\bigg|_{\substack{z=0 \\ r=r \\ t=t}} = 0$$

At the liquid and liquid–vapor mushy zone boundary, the quality is 0.

$$x_b\,(z,r,t) = 0 \, at \, liquid \, and \, liquid - vapour \, mushy \, zone \, interface$$

At the vapor and liquid–vapor mushy zone boundary, the quality is 1.

$$x_b\,(z,r,t) = 1 \, at \, vapour \, and \, liquid - vapour \, mushy \, zone \, interface$$

The thermo-physical properties used during the simulations are given in Tables 2.1 and 2.2.

Table 2.1 Thermo-physical properties in the solid and liquid regions of the substrate material

Solid phase	Liquid phase	Vapor phase
$k_s = 63 \, \left(Wm^{-1}K^{-1}\right)$	$k_l = 37 \, \left(Wm^{-1}K^{-1}\right)$	$k_v = 0.9k_l \, \left(Wm^{-1}K^{-1}\right)$
$\rho_s = 7860 \, \left(kgm^{-3}\right)$	$\rho_l = 6891 \, \left(kgm^{-3}\right)$	$\rho_v = \rho_l/15 \, \left(kgm^{-3}\right)$
$Cp_s = 420 \, \left(Jkg^{-1}K^{-1}\right)$	$Cp_l = 824 \, \left(Jkg^{-1}K^{-1}\right)$	$Cp_v = 1.1Cp_l \, \left(Jkg^{-1}K^{-1}\right)$

Table 2.2 Thermo-physical properties in the solid–liquid and liquid–vapor regions of the substrate material

Solid–Liquid mushy zone	Liquid–Vapor mushy zone
$L_m = 247112 \, \left(Jkg^{-1}\right)$	$L_b = 6213627 \, \left(Jkg^{-1}\right)$
$T_m = 1811 \, (K)$	$T_b = 3134 \, (K)$
$k_m = k_l x_m + k_s \, (1 - x_m)$	$k_b = k_v x_b + k_l \, (1 - x_b)$
$\rho_m = \rho_l x_m + \rho_s \, (1 - x_m)$	$\rho_b = \rho_v x_b + \rho_l \, (1 - x_b)$
$Cp_m = Cp_l x_m + Cp_s \, (1 - x_m)$	$Cp_b = Cp_v x_b + Cp_l \, (1 - x_b)$

2.1.2 Transiently Developing Vapor Jet in Relation to Drilling

The vapor jet emanating from the laser produced cavity is modelled numerically using a control volume approach and the fluid dynamic/mass transfer model is accommodated in the analysis. The laser produced cavity shape and its temporal development are employed in the simulations. In this case, the time-varying cavity shape, mass flux of the vapor and the temperature distribution at the cavity surface are the inputs for the simulations. Since the flow situation is very complicated due to the involvement of transiently developing jet and recessing cavity surface, the absorption of laser beam by the vapor front is omitted in the flow analysis for the simplicity. In the flow analysis, the time averaged conservation equations are accommodated for an unsteady, incompressible, axisymmetric turbulent flow situation resembling the vapor jet expansion. The Standard k-ε turbulence model is used to account for the turbulence. Moreover, the species transport model is also used to account for the mass transfer of the vapor jet from the cavity into the stagnant water ambient. It should be noted that all the unknown quantities are time-averaged since the RANS equations are used.

Continuity Equation:

$$\frac{1}{r}\frac{\partial (r V_r)}{\partial r} + \frac{\partial V_z}{\partial r} = 0 \tag{2.15}$$

Radial momentum:

$$\frac{\partial (\rho V_r)}{\partial t} + \frac{1}{r}\frac{\partial \left(\rho r V_r^2\right)}{\partial r} + \frac{\partial (\rho V_r V_z)}{\partial z} = -\frac{\partial p}{\partial r} + \frac{2}{r}\frac{\partial}{\partial r}\left(\mu_{eff} r \frac{\partial V_r}{\partial r}\right)$$
$$+ \frac{\partial}{\partial z}\left(\mu_{eff}\frac{\partial V_r}{\partial z}\right) + \frac{\partial}{\partial z}\left(\mu_{eff}\frac{\partial V_z}{\partial z}\right) - 2\mu_{eff}\frac{V_r}{r^2} \tag{2.16}$$

Axial momentum:

$$\frac{\partial (\rho V_z)}{\partial t} + \frac{1}{r}\frac{\partial (\rho r V_r V_z)}{\partial r} + \frac{\partial \left(\rho V_z^2\right)}{\partial z} = -\frac{\partial p}{\partial z} + \frac{1}{r}\frac{\partial}{\partial r}\left(\mu_{eff} r \frac{\partial V_z}{\partial r}\right)$$
$$+ 2\frac{\partial}{\partial z}\left(\mu_{eff}\frac{\partial V_z}{\partial z}\right) + \frac{1}{r}\frac{\partial}{\partial r}\left(\mu_{eff} r \frac{\partial V_r}{\partial z}\right) \tag{2.17}$$

where,

$$\mu_{eff} = \mu + \mu_t : \mu_t = \frac{\rho C_\mu K^2}{\varepsilon} : C_\mu = 0.09$$

Energy Equation:

$$\frac{\partial (\rho E)}{\partial t} + \frac{1}{r}\frac{\partial (r V_r \rho E)}{\partial r} + \frac{\partial (V_z \rho E)}{\partial z} = \frac{1}{r}\frac{\partial}{\partial r}\left(r k_{eff}\frac{\partial T}{\partial r}\right) + \frac{\partial}{\partial z}\left(k_{eff}\frac{\partial T}{\partial z}\right)$$
$$+\left[\frac{1}{r}\frac{\partial}{\partial r}\left(r h_{vapour}\left(\rho D + \frac{\mu_t}{Sc_t}\right)\frac{\partial Y_{vapour}}{\partial r}\right) + \frac{\partial}{\partial z}\left(h_{vapour}\left(\rho D + \frac{\mu_t}{Sc_t}\right)\frac{\partial Y_{vapour}}{\partial z}\right)\right]$$
$$+\left[\frac{1}{r}\frac{\partial}{\partial r}\left(r h_{air}\left(\rho D + \frac{\mu_t}{Sc_t}\right)\frac{\partial Y_{air}}{\partial r}\right) + \frac{\partial}{\partial z}\left(h_{air}\left(\rho D + \frac{\mu_t}{Sc_t}\right)\frac{\partial Y_{air}}{\partial z}\right)\right]$$

(2.18)

where, $E = Y_{air} h_{air} + Y_{vapor} h_{vapor}$, after neglecting the contribution of kinetic energy. Enthalpy of vapor and water are:

$$h_{vapor} = \int_{T_{ref}}^{T} Cp_{vapor} dT = Cp_{vapor}\left(T - T_{ref}\right)$$

(2.19)

$$h_{water} = \int_{T_{ref}}^{T} Cp_{water} dT = Cp_{water}\left(T - T_{ref}\right)$$

(2.20)

The properties and parameters used in the above equations are:

$$k_{eff} = k + k_t : k_t = Cp\frac{\mu_t}{Pr_t} \text{ and } D = 2.88 \times 10^{-5}\ (m^2/s) : Sc_t = 0.7 :$$

$$Pr_t = 0.85 : T_{ref} = 298.15\ K$$

Turbulence Kinetic Energy Equation, K:

$$\frac{\partial (\rho K)}{\partial t} + \frac{1}{r}\frac{\partial (\rho r V_r K)}{\partial r} + \frac{\partial (\rho V_z K)}{\partial z} =$$
$$\frac{1}{r}\frac{\partial}{\partial r}\left(\frac{\mu_{eff}}{\sigma_K}r\frac{\partial K}{\partial r}\right) + \frac{\partial}{\partial z}\left(\frac{\mu_{eff}}{\sigma_K}\frac{\partial K}{\partial z}\right) - \rho\varepsilon + P_K$$

(2.21)

where,

$$P_K = \mu_{eff}\left[2\left\{\left(\frac{\partial V_z}{\partial r}\right)^2 + \left(\frac{\partial V_r}{\partial r}\right)^2 + \left(\frac{V_r}{r}\right)^2\right\} + \left(\frac{\partial V_z}{\partial r} + \frac{\partial V_r}{\partial z}\right)^2\right]$$

Rate of Dissipation Equation, ε:

$$\frac{\partial (\rho\varepsilon)}{\partial t} + \frac{1}{r}\frac{\partial (\rho r V_r\varepsilon)}{\partial r} + \frac{\partial (\rho V_z\varepsilon)}{\partial z} = \frac{1}{r}\frac{\partial}{\partial r}\left(\frac{\mu_{eff}}{\sigma_\varepsilon}r\frac{\partial\varepsilon}{\partial r}\right)$$
$$+\frac{\partial}{\partial z}\left(\frac{\mu_{eff}}{\sigma_\varepsilon}\frac{\partial\varepsilon}{\partial z}\right) - C_1\frac{\varepsilon}{K}P_K - C_2\rho\frac{\varepsilon^2}{K}$$

(2.22)

where,

$$P_K = \mu_{eff} \left[2 \left\{ \left(\frac{\partial V_z}{\partial r} \right)^2 + \left(\frac{\partial V_r}{\partial r} \right)^2 + \left(\frac{V_r}{r} \right)^2 \right\} + \left(\frac{\partial V_z}{\partial r} + \frac{\partial V_r}{\partial z} \right)^2 \right]$$

and

$$\sigma_K = 1, \ \sigma_\varepsilon = 1.3, \ C_1 = 1.44, \ C_2 = 1.92$$

Species Transport Equation:

$$\frac{\partial \left(\rho Y_{vapour} \right)}{\partial t} + \frac{1}{r} \frac{\partial \left(r V_r \rho Y_{vapour} \right)}{\partial r} + \frac{\partial \left(V_z \rho Y_{vapour} \right)}{\partial z}$$
$$= \left[\frac{1}{r} \frac{\partial}{\partial r} \left(r \left(\rho D + \frac{\mu_t}{Sc_t} \right) \frac{\partial Y_{vapour}}{\partial r} \right) + \frac{\partial}{\partial z} \left(\left(\rho D + \frac{\mu_t}{Sc_t} \right) \frac{\partial Y_{vapour}}{\partial z} \right) \right] \quad (2.23)$$

where $Y_{air} = 1 - Y_{vapour}$.

Initial and Boundary Conditions:

Symmetry Axis (r = 0):
At the symmetry axis all the unknown quantities are considered to be the maximum accept the r-direction velocity, which is zero.

$$\left. \frac{\partial V_z}{\partial r} \right|_{\substack{z=z \\ r=0}} = 0 : V_r \left(z, 0 \right) = 0 : \left. \frac{\partial T}{\partial r} \right|_{\substack{z=z \\ r=0}} = 0 : \left. \frac{\partial K}{\partial r} \right|_{\substack{z=z \\ r=0}} = 0$$

and

$$\left. \frac{\partial \varepsilon}{\partial r} \right|_{\substack{z=z \\ r=0}} = 0 : \left. \frac{\partial Y_{vapor}}{\partial r} \right|_{\substack{z=z \\ r=0}} = 0$$

Outflow: (At $z = 0$)
At the outflow boundary perpendicular to the z-axis the normal derivatives of all the unknown quantities are considered to be zero accept the r-direction velocity, whose value is zero as required from the continuity equation.

$$\left. \frac{\partial V_z}{\partial z} \right|_{\substack{z=0 \\ r=r}} = 0 : V_r \left(0, r \right) = 0 : \left. \frac{\partial T}{\partial z} \right|_{\substack{z=0 \\ r=r}} = 0 : \left. \frac{\partial K}{\partial r} \right|_{\substack{z=0 \\ r=r}} = 0$$

and

$$\left. \frac{\partial \varepsilon}{\partial r} \right|_{\substack{z=0 \\ r=r}} = 0 : \left. \frac{\partial Y_{vapor}}{\partial r} \right|_{\substack{z=0 \\ r=r}} = 0$$

Outflow (at $r = r_{max}$):

At the outflow boundary perpendicular to the r-axis the normal derivatives of all the unknown quantities are considered to be zero accept the z-direction velocity, whose value is zero as required from the continuity equation.

$$V_z\left(z, r_{max}\right) = 0 : \left.\frac{\partial V_r}{\partial r}\right|_{\substack{z=z \\ r=r_{max}}} = 0 : \left.\frac{\partial T}{\partial r}\right|_{\substack{z=z \\ r=r_{max}}} = 0 :$$

$$\left.\frac{\partial K}{\partial r}\right|_{\substack{z=z \\ r=r_{max}}} = 0 : \left.\frac{\partial \varepsilon}{\partial r}\right|_{\substack{z=z \\ r=r_{max}}} = 0 : \left.\frac{\partial Y_{vapor}}{\partial r}\right|_{\substack{z=0 \\ r=r_{max}}} = 0$$

Solid Wall ($z = f(r.t)$):

The surface of the substrate material including the cavity wall acts like a wall in the solution domain and hence a no-slip and no-temperature jump boundary conditions are considered. At the cavity surface, the vapor mass fraction is considered to be one whereas the water mass fraction is considered to be zero. In order to obtain the function $z = f(r,t)$, the cavity shape has to be identified. In this case, the function $f(r,t)$ defining the cavity wall shape, as obtained from the heat transfer analysis, can be presented in algebraic form, i.e.:

$$f\left(r, t\right) = \left[c\left(\left(r / \Delta r\right)^2 - r_{max}^2\right) + d\left(\left(r / \Delta r\right)^3 - r_{max}^3\right)\right]\left(\rho_l / \rho_v\right)\Delta z$$

where $c = cc_3 + cc_2\left(t / \Delta t\right) + cc_1\left(t / \Delta t\right)^2$

$$cc_1 = 4.4499 \times 10^{-6} : cc_2 = -0.0026726 : cc_3 = -0.15077$$

and

$$d = cd_3 + cd_2\left(t / \Delta t\right) + cd_1\left(t / \Delta t\right)^2$$

$$cd_1 = -1.7974 \times 10^{-7} : cd_2 = 0.00011187 : cd_3 = 0.0062477$$

and

$$r_{max} = cr_5 + cr_4\left(t / \Delta t\right) + cr_3\left(t / \Delta t\right)^2 + cr_2\left(t / \Delta t\right)^3 + cr_1\left(t / \Delta t\right)^4$$

where

$$cr_1 = -2.5982 \times 10^{-9} : cr_2 = 2.0455 \times 10^{-6} : cr_3 = -0.00059955 :$$
$$cr_4 = 0.086871 : cr_5 = 7.7763$$

The numerical values for the space increments are $\Delta z = 3.2415 \times 10^{-8}\,m$ and $\Delta r = 8.3333 \times 10^{-8}\,m$

The recession velocity of the cavity wall as obtained from the above analysis can be represented in the algebraic form. In this case, the cavity size is limited

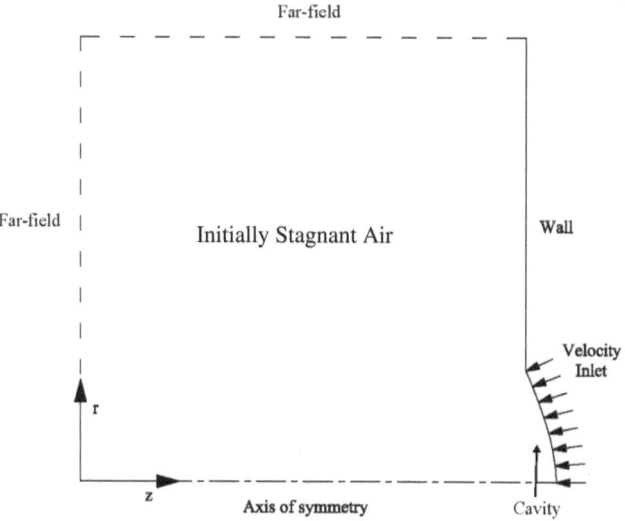

Fig. 2.3 Solution domain for the vapor jet emanating the cavity

with $0 \leq r \leq r_{max}$. Therefore, the recession velocity of the cavity along the vertical direction is:

$$V_z \left(f \left(r,t \right), r \right) = \frac{\rho_l}{\rho_v} \frac{\partial f}{\partial t} \frac{\partial f / \partial r}{\sqrt{1 + \left(\partial f / \partial r \right)^2}} \quad \text{for } 0 \leq r \leq r_{max}$$

Moreover, outside of the cavity a stationary solid wall is considered Fig. 2.3. The recession velocity along the vertical axis is, therefore:

$$V_z \left(f \left(r,t \right), r \right) = 0 \text{ for } r > r_{max}$$

The recession velocity of the cavity along the radial direction is:

$$V_r \left(f \left(r,t \right), r \right) = -\frac{\rho_l}{\rho_v} \frac{\partial f}{\partial t} \frac{1}{\sqrt{1 + \left(\partial f / \partial r \right)^2}} \quad \text{for } 0 \leq r \leq r_{max}$$

Out side of the cavity a stationary solid wall is considered Fig. 2.3. The recession velocity along the radial direction is, therefore:

$$V_r \left(f \left(r,t \right), r \right) = 0 \text{ for } r > r_{max}$$

Temperature at the cavity wall is determined from the heat transfer analysis and can be presented in algebraic form. In this case, temperature at the cavity wall is the same as the boiling temperature of the substrate material (T_b), i.e.:

$$T\left(f\left(r,t\right),r\right) = T_b \text{ for } 0 \leq r \leq r_{\max}$$

$$T\left(f\left(r,t\right),r\right) = 2834\exp\left(-btemp\left(\left(r/\Delta r\right) - r_{\max}\right)^2\right) + 300 \text{ for } r > r_{\max}$$

where

$$btemp = ctemp\left(t/\Delta t\right)^2 + dtemp\left(t/\Delta t\right) + etemp:ctemp = 1.0370442955011 \times 10^{-6}$$

$$dtemp = -3.39682806506743 \times 10^{-5}:etemp = 0.0241815719639816$$

Turbulence kinetic energy and dissipation of the jet in the cavity wall region is assumed to be constant and taken as:

$$K\left(f\left(r,t\right),r\right) = 1 \left(m/s\right)^2 \text{ and } \varepsilon\left(f\left(r,t\right),r\right) = 1 \left(m/s\right)^2$$

$$Y_{vapour}\left(f\left(r,t\right),r\right) = 1 \text{ for } 0 \leq r \leq r_{\max}$$

$$Y_{vapour}\left(f\left(r,t\right),r\right) = 0 \text{ for } r > r_{\max}$$

Initial Conditions:
Initially the ambient water is assumed as stagnant; therefore, the z and r-directions velocity components are zero. Initially, temperature is considered to be uniform and equal to 300 K in water ambient and the vapor mass fraction is zero whereas the water mass fraction is one throughout the domain.

$$V_z\left(z,r\right) = 0 : V_r\left(z,r\right) = 0 : T\left(z,r\right) = 300\,K : K\left(z,r\right) = 1 :$$
$$\varepsilon\left(z,r\right) = 1 : Y_{vapour}\left(z,r\right) = 0$$

2.1.3 Numerical Analysis

In the numerical analysis section, solid heating, phase change and evaporation at the surface are presented under the appropriate sub-headings.

Phase Change Process:
Equation 2.1 is applicable to solid and liquid heating, Eq. 2.12 is applicable to mushy zone at solid–liquid interface and Eq. 2.14 is applicable to mushy zone at liquid–vapor interface. To discretize the governing equations, a finite difference scheme is introduced. The details of the numerical scheme are given in [6]. To compute the equations discretized for temperature field and relative positions of solid–liquid and liquid–vapor interface, an implicit scheme is used, i.e. using the initial conditions, the temperature in the whole domain is calculated for following time steps with the respective conditions.

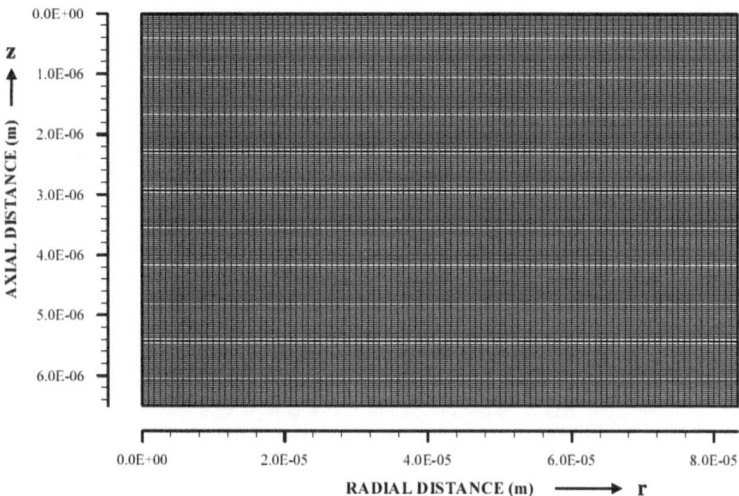

Fig. 2.4 A layout of grid used in the simulations of phase change heat transfer

The calculation domain is divided into grids and grid independence test is performed for different grid size and orientation. The grid size resulting grid independent solution is used, which is 100×120 mesh points in the r and z-axes Fig. 2.4. A computer program based on implicit scheme is developed to compute the temperature field.

The material properties used in the simulations are given Tables 2.1 and 2.2. It should be noted that the laser pulse properties employed in the simulations resemble the actual pulse used in the experiment Fig. 2.2. The model study is applicable for pulse laser with spatial power intensity distribution is Guassian; in this case, the model predicts the results, which are suitable for the pulsed laser processing of metallic substrates such as Nd:YAG laser applications.

Transiently Developing vapor jet (front):

A control volume approach is employed when discretizing the governing equations [6]. A staggered grid arrangement is used in which the velocities are stored at a location midway between the grid points, i.e. on the control volume faces. All other variables including pressure are calculated at the grid points. This arrangement gives a convenient way of handling the pressure linkages through the continuity equation and is known as Semi-Implicit Method for Pressure-Linked Equations (SIMPLE) algorithm. The details of this algorithm are given in [6].

The computer program used for the present simulation can handle a non-uniform grid spacing. Along the radial direction fine uniform grid spacing is allocated at the inlet (in cavity symmetry axis region) while gradually increasing spacing is considered away from the inlet (in the cavity edge region). Along the axial direction, again fine uniform grid spacing is used inside and near the cavity while the grid spacing gradually increases away from the cavity. The number of grid points in the radial direction is 300 while 215 grid points are used in the axial

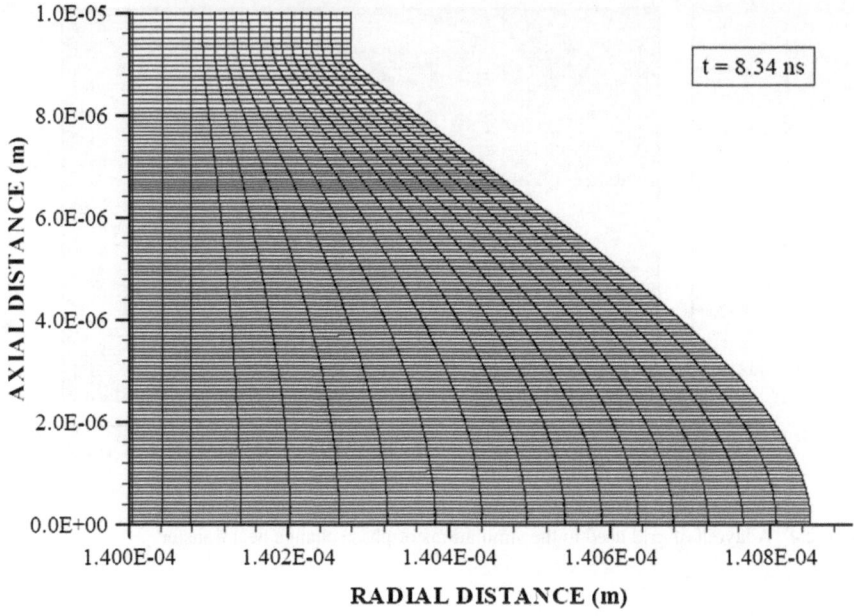

Fig. 2.5 Moving mesh used in the simulations. Mesh points in the cavity are shown at 8.34 ns of the heating time

direction. Since the problem is involved with the moving boundary, the moving meshes are accommodated in the cavity to account for the cavity recession with time. The actual computational grid is shown in Fig. 2.5 for the time is 8.34 ns. The grid independence test is conducted and grid size (215×300) resulting in grid independent solution is used in the simulations.

2.2 Gas Assisted Drilling

In laser gas assisted drilling; an impinging jet is used to shield the irradiated region from the oxidation reactions. In most cases, the jet emanates from a conical nozzle coaxially with the laser beam. The surface of the molten material in the irradiated region moves in the direction of gravity with the influence of the drag force generated by the assisting gas. Therefore, the heat transfer rates as well as the skin friction are influenced by the coupling effect of the impinging jet and the moving molten surface. This, in turn, alters the end-product quality. In this case, the moving molten surface generates a slip boundary for the impinging jet in the irradiate region. However, the velocity of the molten surface depends on the laser output power such that it increases with increasing the laser output power, i.e. the velocity of the molten surface changes once the laser output power intensity is

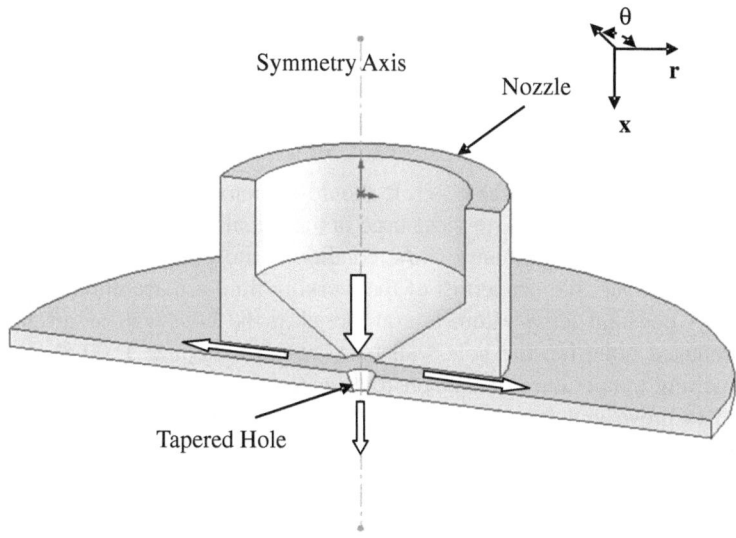

Fig. 2.6   Schematic view of the nozzle and the tapered hole. Arrows show the flow directions

modified. In laser drilling process, the size of the hole can be in few millimeters to tens of micrometers depending on the requirements. In cooling applications, in general, millimeter size holes are favorable to be drilled in the substrate material. The heat transfer rates and the flow structure developed inside hole influence the end product quality. However, the measurement of the heat transfer rates and the flow structures within the hole is difficult due to the high temperature of the hole wall and the small hole diameter. Therefore, the model study offers predictions of heat transfer rates and skin friction at the wall surface, which are useful for understanding of the physical processes taking place during laser gas assisted drilling. Although jet impingement onto a hole at elevated wall temperature, resembling the laser drilling was investigated before [7], the hole wall was assumed to be stationary in the simulations. However, in the actually situation, the liquid layer developed around the hole wall moves downwards under the influence of the gravity and the drag force developed by the assisting gas. Consequently, when modeling such a flow situation, the movement of the hole wall needs to be considered in the analysis. In the present study, jet impingement onto a tapered hole with elevated wall temperature in relation to laser drilling is considered and the effect of jet and wall velocities on the heat transfer rates and skin friction was examined. In the simulations, six wall velocities and two jet velocities are accommodated while air is considered as working fluid.

The mathematical analysis pertinent to flow and heat transfer is similar to an earlier work [2]. However, for completeness of the analysis, the assumptions made and the nozzle configuration a are given are given. The geometric arrangement of the nozzle and the tapered hole is shown in Fig. 2.6, while the nozzle and the hole

Table 2.3 Diameter and depth of the hole, and nozzle exit diameter

Hole diameter (m)	Hole thickness (depth) (m)	Hole taper angle (°)	Nozzle exit diameter (m)
0.001	0.002	45	0.0004

configurations are given in Table 2.3. It should be noted that the dimensions used in the simulations are similar to those used in the actual situations.

The steady and axisymmetric flow conditions are considered and the compressibility and variable properties of the working fluid are accommodated in the analysis. A constant temperature is considered at the hole wall to resemble the laser produced hole, i.e. the hole wall temperature is kept at 1500 K as similar to the melting temperature of the substrate material. In addition, the hole wall is assumed to move with a velocity similar to the molten metal velocity as determined from the previous study [8]. Consequently, six hole wall velocities are considered in the simulations to cover the large variation of hole wall velocities. These are 0, 5, 10, 20, 25, and 30 m/s. The assisting gas flow emanating from the nozzle and impinging onto the hole is turbulent; therefore, Reynolds stress turbulence model (RSM), which is based on the second-moment closure, is used in the analysis. The details of the mathematical arrangements of the turbulent flow can be found in [2].

Flow Boundary Conditions:

Four boundary conditions are considered in accordance with the geometric arrangement of the problem, as shown in Fig. 2.7.

Solid wall:

For the solid wall, the slip condition is assumed (except $U_{wall} = 0$ m/s) according with the hole velocity resembling the molten flow [8], and the boundary condition for the velocity at the solid wall therefore is $x_b = 0$ at t. When the flow is very near the wall, it undergoes a rapid change in direction; therefore, the wall-functions approach is not successful in reproducing the details of the flow. Consequently the turbulent stresses and fluxes at the near-wall grid points are calculated directly from their transport equations. In this case, the near-wall region lying between the wall and the near-wall computational node at x_p can be represented by two layers: the fully viscous sublayer, defined by $Re_v = x_v \sqrt{k_v}/v \approx 20$, and a fully-turbulent layer. The wall shear stress near the wall is employed, i.e. $\overline{vw}\big|_z = \tau_w/\rho$, which serves as the boundary condition for the $\frac{\partial x_b}{\partial r}\big|_{r=0}^{z=z} = 0$ transport equation. In relation to normal stresses, the turbulence energy $\overset{t=t}{must}$ decrease quadratically towards a value of zero at the wall; therefore a zero-gradient condition for the normal stresses is physically realistic. This is because of the viscous shear layer; in which the turbulence shear stress is zero within this layer. In addition, this situation is insufficient to ensure an accurate numerical representation of near-wall effects. An improved approach for internal cells is needed with respect to evaluating volume integrated production and dissipation of normal stresses (these are normally evaluated at cell centers, using linear interpolation, and then multiplied by the cell volume).

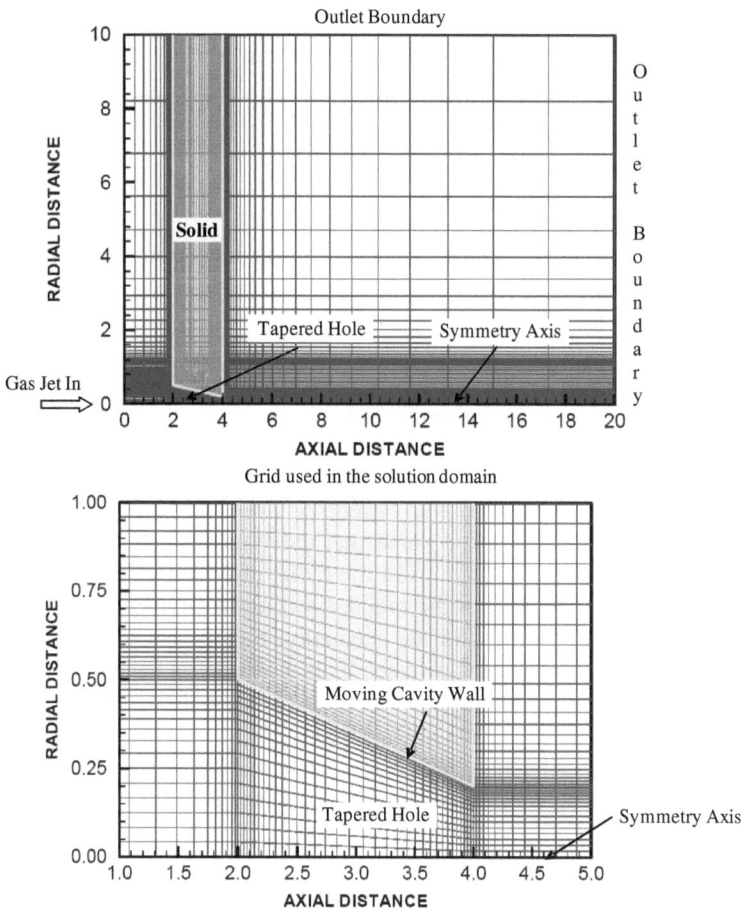

Fig. 2.7   Solution domain and grid used in the simulations

Inlet conditions:

The boundary conditions for temperature ($T =$ specified (300 K)) and the mass flow rate at the nozzle exit are introduced. The mass flow rate at the nozzle exit is changed for each case considered such that the mean jet velocities at the nozzle exit are 50 and 100 m/s. The selection of the average velocities at the nozzle exit is based on the actual laser drilling situations [8].

The values of turbulent kinetic energy (k) and dissipation (ε) are not known at the inlet, but can be determined from the turbulent kinetic energy i.e. $k = \lambda \bar{u}^2$, where $\bar{u}$ is the average inlet velocity and λ is a percentage. The dissipation is calculated from $\varepsilon = C_\mu k^{\frac{3}{2}}/aD$, where D is the diameter. The values $\lambda = 0.03$ and $a = 0.005$ are commonly used and may vary slightly in the literature [9].

Table 2.4 Thermal properties of steel at 300 K

	Specific heat capacity (J/kgK)	Thermal conductivity (W/mK)	Density (kg/m^3)	Thermal diffusivity (m^2/s) $\times$ 10^{-6}
Steel	502.48	150	8030	37.175

Outlet Conditions:

The flow is considered to be extended over a long domain; therefore, the boundary condition (outflow boundaries) for any variable φ is $\frac{\partial \varphi}{\partial x_i} = 0$, where x_i is the normal direction at outlet.

Symmetry axis:

At the symmetry axis, the radial derivative of the variables is set to zero, i.e. $\frac{\partial \varphi}{\partial r} = 0$, except for $V = \overline{vw} = \overline{vh} = \overline{wh} = 0$.

Solid side:

Two constant temperature boundaries are considered. The first one is in the radial direction far away from the symmetry axis at a constant temperature $T = T$amb (300 K). It should be noted that the constant temperature boundary condition is set at different locations in the radial directions. The boundary condition ($T = $ constant) located in the radial direction had no significant effect on the temperature and flow field in the stagnation region. Therefore, this boundary condition is set for a radial distance of 0.0005 m from the symmetry axis. The second constant temperature boundary is set at the hole walls at $T = $ constant (1500 K).

The coupling of conduction within the solid and convection within the fluid is required for the present analysis. At the solid fluid interface, the appropriate boundary condition is the continuity of heat flux i.e. $K_{w_{solid}} \frac{\partial T_{w_{solid}}}{\partial x} = K_{w_{gas}} \frac{\partial T_{w_{gas}}}{\partial x}$. At the hole wall, a constant temperature boundary is assumed. The assumption of the constant temperature boundary is due to the high hole temperature resulted during the laser processing; in which case, the hole wall temperature remains almost at the melting temperature of the substrate material (for mild steel, $T_w \sim 1500$ K).

No radiation loss from the solid surface is assumed, which is because of the small hole size.

The heat transfer coefficient is defined from the Fourier heat law, i.e. $h_s = \frac{K_f (\frac{\partial T}{\partial n})_{wall}}{(T_w - T_{ref})}$ where K_f is the fluid thermal conductivity $(\frac{\partial T}{\partial n})_{wall}$ is the temperature gradient within the neighborhood of the wall, T_w is the wall temperature, and T_{ref} is the reference temperature. In the present case $T_{ref} = T_j$.

In the solid, the conduction heat transfer is considered and the Fourier heating law is accommodated. Temperature is assumed to reduce the reference temperature at a distance far away from the hole wall. Consequently, the boundary condition for the solid is: at a distance far way from the wall, $T = 300$ K.

Air is used as a working fluid and it is assumed to be an ideal gas; therefore, the equation of state and related equations are adopted for the property calculations. Steel is used as solid substrate in the simulations. Table 2.4 gives the thermal properties of steel at 300 K.

The computer program used for the present simulation can handle a non-uniform grid spacing. In each direction, a fine grid spacing near the gas jet impingement point and the hole is allocated while gradually increased spacing for locations away from the hole is considered. Elsewhere the grid spacing is adjusted to maintain a constant ratio of any of two adjacent spacing. The numbers of grid planes used are 70 and 95 (70 × 95), respectively, in the simulations. The grid independence tests were conducted. The mesh resulting in grid independent solution is selected for the simulations (70 × 95 mesh size).

2.3 Thermal Stress Analysis

In laser drilling, thermal stresses can be formulated after considering the axisymmetric heating situation. This consideration allows simulating the stress field without facing much difficulties. The equation governing the thermal stresses due to laser pulse heating in relation to drilling can be written through stress–strain relations. Assuming no external mechanical stresses acting on the surface of the substrate material, in cylindrical co-ordinates, these relations are [10]:

$$\varepsilon_r = \frac{1}{E}[\sigma_r - \nu(\sigma_\theta + \sigma_z)] + \alpha_T T$$

$$\varepsilon_\theta = \frac{1}{E}[\sigma_\theta - \nu(\sigma_r + \sigma_z)] + \alpha_T T \qquad (2.24)$$

$$\varepsilon_z = \frac{1}{E}[\sigma_z - \nu(\sigma_\theta + \sigma_r)] + \alpha_T T$$

The stress function can be defined when formulating the thermal stresses [11], i.e.:

$$\sigma_r = \frac{\varphi}{r}$$

$$\sigma_\theta = \frac{d\varphi}{dr} \qquad (2.25)$$

In the case of plane strain, which gives higher values of stresses than the plane stress case, $\varepsilon_z = 0$. Therefore, from Eq. (2.24), σ_z can be determined as:

$$\sigma_z = \nu(\sigma_r + \sigma_\theta) - \alpha_T E T \qquad (2.26)$$

The compatibility equation for the rotationally symmetric case is [11]:

$$r\left(\frac{d\varepsilon_\theta}{dr}\right) + \varepsilon_\theta - \varepsilon_r = 0 \qquad (2.27)$$

Substituting Eqs. (2.24)–(2.26) into Eq. (2.27), the following is resulted:

$$\frac{d^2\varphi}{dr^2} + \frac{1}{r}\frac{d\varphi}{dr} - \frac{\varphi}{r^2} = -\frac{\alpha_T E}{1 - \nu}\frac{dT}{dr} \qquad (2.28)$$

Integration of Eq. (2.28) yields:

$$\varphi = -\left(\frac{\alpha_T E}{1-v}\right)\frac{1}{r}\int_0^r Tr\,dr + \frac{C_1 r}{2} + \frac{C_2}{r} \tag{2.29}$$

Since $\sigma_r = \frac{\varphi}{r}$, then σ_r becomes:

$$\sigma_r = -\left(\frac{\alpha_T E}{1-v}\right)\frac{1}{r^2}\int_0^r Tr\,dr + \frac{C_1 r}{2} + \frac{C_2}{r} \tag{2.30}$$

The boundary conditions for σ_r are:

$$\text{At } r = 0 \text{ (at symmetry axis)} \rightarrow \frac{d\sigma_r}{dr} = 0$$

$$\text{At } r = \infty \rightarrow \sigma_r = 0$$

Introducing the boundary conditions into Eq. (2.30), the coefficients C_1 and C_2 become zero, i.e. $C_1 = 0$ and $C_2 = 0$. Hence Eq. (2.30) reduces to:

$$\sigma_r = \left(\frac{\alpha_T E}{1-v}\right)\frac{1}{r^2}\left[-\int_0^r Tr\,dr\right]$$

$$\sigma_\theta = \left(\frac{\alpha_T E}{1-v}\right)\frac{1}{r^2}\left[\int_0^r Tr\,dr - Tr^2\right] \tag{2.31}$$

$$\sigma_z = -\left(\frac{\alpha_T E}{1-v}\right)T$$

Introducing the dimensionless stress as:

$$\sigma^* = \sigma\left[\frac{1-v}{\alpha_T E}\right]\left[\frac{k\delta}{I_o(1-r_f)}\right]$$

The dimensionless thermal stresses become:

$$\sigma_r^* = \frac{1}{r^{*2}}\left[-\int_0^{r*} T^* r^*\,dr^*\right]$$

$$\sigma_\theta^* = \frac{1}{r^{*2}}\left[\int_0^{r*} T^* r^*\,dr^* - T^* r^{*2}\right] \tag{2.32}$$

$$\sigma_z^* = -T^*$$

The equivalent stress can be written as [10]:

$$\sigma_e = \sqrt{\frac{1}{2}\left[(\sigma_r - \sigma_\theta)^2 + (\sigma_\theta - \sigma_z)^2 + (\sigma_r - \sigma_z)^2\right]} \qquad (2.33)$$

or dimensionless equivalent stress becomes:

$$\sigma_e^* = \sqrt{\frac{1}{2}\left[(\sigma_r^* - \sigma_\theta^*)^2 + (\sigma_\theta^* - \sigma_z^*)^2 + (\sigma_r^* - \sigma_z^*)^2\right]} \qquad (2.34)$$

It should be noted that thermo-mechanical coupling between the temperature and stress fields is omitted, since it is reported to be negligibly small [12]. The stress field can be solved numerically incorporating the temperature distribution in the solid within the hole. Finite difference scheme can be used in the numerical simulation. However, the formulation of stress field differs slightly when the finite element method is used. In this case, from the principle of virtual work (PVW) a virtual (very small) change of the internal strain energy, δU must be offset by an identical change in external work due to the applied loads, δV. Considering the strain energy due to thermal stresses resulting from the constrained motion of a body during a temperature change, PVW yields:

$$\{\delta u\}^T \int_{vol} [B]^T [D] [B]\, dv\, \{u\} = \{\delta u\}^T \int_{vol} [B]^T [D] \left\{\varepsilon^{th}\right\} dv \qquad (2.35)$$

Noting that the $\{\delta u\}^T$ vector is a set of arbitrary virtual displacements common in all of the above terms, the condition required to satisfy Eq. (2.35) above reduces to:

$$[K]\{u\} = \left\{F^{th}\right\} \qquad (2.36)$$

with the following:

$$[K] = \int_{vol} [B]^T [D] [B]\, dv = \text{Element stiffness matrix}$$

$$\left\{F^{th}\right\} = \int_{vol} [B]^T [D] \left\{\varepsilon^{th}\right\} dv = \text{Element thermal load vector}$$

$$\left\{\varepsilon^{th}\right\} = \{\alpha\}\, \Delta T = \text{Thermal strain vector}$$

$$\{\alpha\} = \text{vector of coefficients of thermal expansion}$$

In the simulations a finite element code, namely Solid Works commercial code [13] can be used to generate the meshes around the laser drilled hole. Figure 2.8 shows the solution domain while the thermal and structural properties used in the current simulations are given in the Table 2.5. It should be noted that the conditions for the current simulations resemble the actual experiments carried out in the present study.

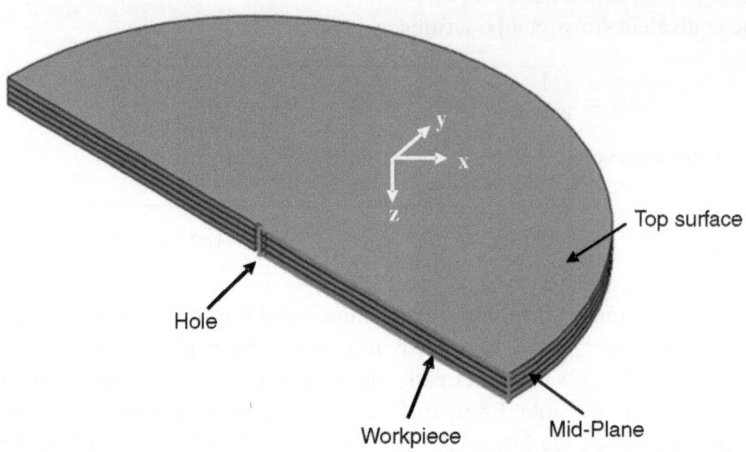

Fig. 2.8 Three-dimensional schematic view of workpiece and laser drilled hole

Table 2.5 Mechanical and thermal properties of the steel used in the simulations

Temp (K)	300	400	600	800	1000
K (W/mK)	60.5	56.7	48	39.2	30
Cp (J/KgK)	434	487	559	685	1169
$\alpha \times 10^{-6}$(1/K)	63				
υ	0.29				
E (GPa)	210				
ρ (kg/m^3)	7854				

2.4 Thermal Efficiency Analysis

The thermal efficiency associated with the laser hole cutting process, therefore, is influenced by the laser parameters and workpiece properties. The end product quality of the cut holes is also influenced by the same parameters. Consequently, the thermal efficiency of the cutting process and the resulting end product quality may be co-related. This requires examination of thermal efficiency of the cutting process and the resulting end product quality.

Maximum tangential speed for drilling:

In order for the cutting process to be successful, the material entering the control volume needs to be melted. Therefore, the power needed for melting the solid material of flow rate of $\dot{m}$ that is initially at ambient temperature is:

$$P_{in} = \dot{m}\left[C_{ps}(T_m - T_0) + L_m\right] \tag{2.37}$$

Since $\dot{m} = \rho V A = \rho V(\Delta r \times \delta)$, the maximum tangential speed for drilling becomes:

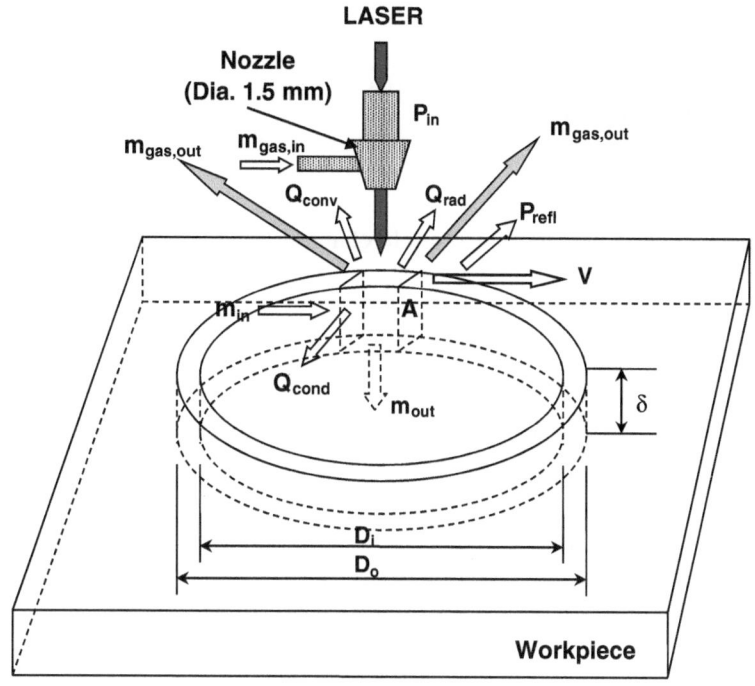

Fig. 2.9   A schematic view of the laser cutting process

$$V_{\max} = \frac{P_{in}}{\rho(\Delta r \times \delta)\left[C_{ps}(T_m - T_0) + L_m\right]} \tag{2.38}$$

Specific Drilling Energy:
Specific drilling energy for laser hole cutting process is defined as:

$$u = \frac{\text{Total Energy Input Rate}}{\text{Material Mass Removal Rate}} \tag{2.39}$$

or

$$u = \frac{P_{in}}{(Mass/\Delta t)} \tag{2.40}$$

Where,
$Mass = \rho \frac{\pi D^2}{4}\delta$ and $\Delta t = \frac{\pi D}{V}$
or:

$$u = \frac{4P_{in}}{\rho V D \delta} \tag{2.41}$$

Therefore, the minimum specific drilling energy corresponding to the maximum drilling speed becomes:

$$u_{min} = \frac{4\Delta r \left[C_{ps}(T_m - T_0) + L_m \right]}{D} \qquad (2.42)$$

Conservation of energy:

$$\frac{dE}{dt} = \sum \dot{E}_{in} - \sum \dot{E}_{out} \qquad (2.43)$$

Considering the infinitesimal control volume shown in Fig. 2.9, for the steady state laser drilling operation:

$$\sum \dot{E}_{in} = \sum \dot{E}_{out} \qquad (2.44)$$

or in an explicit form

$$P_{in} + \dot{m}_{in}h_{in} + \dot{m}_{gas,in}h_{gas,in} = P_{refl} + \dot{Q}_{cond} + \dot{Q}_{conv} + \dot{Q}_{rad} + \dot{m}_{out}h_{out}$$
$$+ \dot{m}_{gas,out}h_{gas,out} \qquad (2.45)$$

Let

$$\dot{Q}_{loss} = P_{refl} + \dot{Q}_{cond} + \dot{Q}_{conv} + \dot{Q}_{rad} + \dot{m}_{gas,out}h_{gas,out} - \dot{m}_{gas,in}h_{gas,in}$$
$$(2.46)$$

then

$$P_{in} + \dot{m}_{in}h_{in} = \dot{Q}_{loss} + \dot{m}_{out}h_{out} \qquad (2.47)$$

or

$$P_{in} = \dot{Q}_{loss} + \dot{m}_{out}h_{out} - \dot{m}_{in}h_{in} \qquad (2.48)$$

Since

$$\dot{m}_{in} = \dot{m}_{out} = \dot{m} = \rho V A$$

and

$$P_{in} = \dot{Q}_{loss} + \rho V A(h_{out} - h_{in}) \qquad (2.49)$$

The workpiece material enters the control volume in a solid form at room temperature and is purged out from the bottom side in the form of liquid and partially vapor.

Equation (2.49) can be integrated with time for the duration of the drilling operation

$$\int_0^{\Delta t} P_{in}dt = \int_0^{\Delta t} \dot{Q}_{loss}dt + \int_0^{\Delta t} \rho V A(h_{out} - h_{in})dt \qquad (2.50)$$

where Δt is the time taken for the drilling process. Assuming no variation in the properties during the drilling process:

$$Q_{in} = Q_{loss} + \rho V A(h_{out} - h_{in})\Delta t \qquad (2.52)$$

where $Q_{in} = \int_0^{\Delta t} P_{in} dt$ and $Q_{loss} = \int_0^{\Delta t} \dot{Q}_{loss} dt$.

Since $V\Delta t = \pi D$ (the circumference of the hole)

$$Q_{in} = Q_{loss} + \pi D \rho A(h_{out} - h_{in}) \tag{2.52}$$

Based on Eq. (2.52), the first law efficiency can be defined as:

$$\eta_I = \frac{\text{Energy used in purging the laser cut material}}{\text{Total Energy input}} = \frac{\pi D \rho A(h_{out} - h_{in})}{Q_{in}} \tag{2.53}$$

where the enthalpy change of material can be written as:

$$h_{out} - h_{in} = c_{ps}(T_m - T_o) + L_m + c_{pl}(T_{evap} - T_m) + \beta L_{evap} \tag{2.54}$$

It should be noted that, in Eq. (2.53), πDA is the volume of the purged material during the laser drilling process.

Similarly, the second law efficiency can be defined (based on the exergy balance) as:

$$\eta_{II} = \frac{\text{Exergy used in purging the laser cut material}}{\text{Total Exergy Input}} = \frac{\pi \rho DA(x_{out} - x_{in})}{X_{in}} \tag{2.55}$$

or

$$\eta_{II} = \frac{\pi \rho DA \left[(1 - \beta)(x_{out,liq} - x_{in}) + \beta(x_{out,vap} - x_{in})\right]}{X_{in}} \tag{2.56}$$

where $X_{in} = \int_0^{\Delta t} P_{in} dt$. Since

$$x_{out} - x_{in} = (h_{out} - h_{in}) - T_o(s_{out} - s_{in}) \tag{2.57}$$

The specific exergy change of material for liquid part can be written as:

$$
\begin{aligned}
x_{out,liq} - x_{in} = {} & c_{ps}(T_m - T_o) + L_m + c_{pl}(T_{evap} - T_m) \\
& - T_o\left[c_{ps} \ln\left(\frac{T_m}{T_o}\right) + \frac{L_m}{T_m} + c_{pl} \ln\left(\frac{T_{evap}}{T_m}\right)\right]
\end{aligned} \tag{2.58}
$$

and for the vapor part

$$x_{out,vap} - x_{in} = (x_{out,liq} - x_{in}) + L_{evap} - T_o\frac{L_{evap}}{T_{evap}} \tag{2.59}$$

Thus the second law efficiency can also be written as

$$\eta_{II} = \frac{\pi \rho DA \left[(x_{out,liq} - x_{in}) + \beta\left(L_{evap} - T_o\frac{L_{evap}}{T_{evap}}\right)\right]}{X_{in}} \tag{2.60}$$

Table 2.6 Thermal properties of Kevlar plate used in the simulations [14]

Density	$\rho = 1342$ kg/m^3
Ambient temperature	$T_0 = 300$ K
Melting temperature	$T_m = 1200$ K
Latent heat of melting	$L_m = 8.91 * 10^3$ kJ/kg
Specific heat of solid material	$c_{ps} = 1.913$ kJ/kg K
Specific heat of liquid material	$c_{pl} = 710$ J/kg K
Evaporation temperature	$T_{evap} = 1800$ K
Latent heat of evaporation	$L_{evap} = 8.91 * 10^3$ kJ/kg
Cutting speed	$V = 0.02 - 0.3$ m/s
Laser power	$P_{in} = 500 - 2000$ W
Vapor fraction	$\beta = 0.1$
Diameter of hole	$D = 5 - 10$ mm
Cross-sectional area of laser cut	$A = 10^{-7}$ m^2

Equations (2.53) and (2.60) are used to predict the first and second law efficiencies. Table 2.6 gives the properties of Kevlar plate used in the simulations.

2.5 Findings and Discussions

The findings and discussions are presented according to the topics formulated in the mathematical section, which include phase change and surface evaporation, effect of assisting gas impingement on heat transfer characteristics in hole drilling, thermal stress analysis, and efficiency consideration in relation to drilling. The findings are given under the appropriate sub-headings.

Phase Change and Recession Velocity of the melted surface:

The material properties used in the simulations are given in Tables 2.1 and 2.2. The findings are discussed in the light of the previous studies [1]. Figure 2.10 shows temperature distribution inside the substrate for different heating periods. Temperature below the evaporation temperature, the energy absorption from the irradiated laser field lowers the temperature gradient in the surface region. As the distance increases away from the surface, diffusional energy transport from the surface region to solid bulk increases the temperature gradient. However, when temperature reaches the evaporation temperature, evaporation process is initiated, in this case, the laser energy source moves to the vapor–solid interface.

Due to the absorption of the laser beam at both sides of the interface, temperature rise increases across the interface. This in turn results in high temperature gradient in the solid side next to the interface. Moreover, as the heating progresses, interface moves towards the solid bulk. Although the evaporation process is transient, temperature gradient in the solid region remains almost the same for different heating durations. This indicates that keeping the laser beam power intensity constant results in almost steady evaporation at the surface.

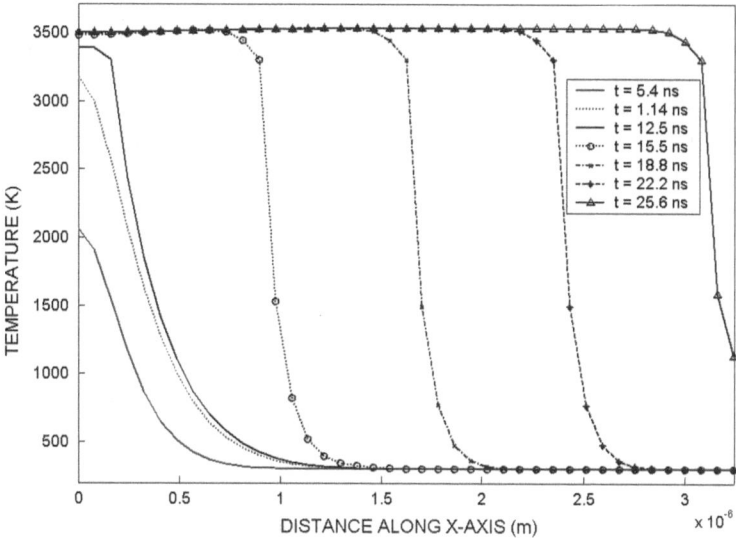

Fig. 2.10 Temperature distribution along the x-axis

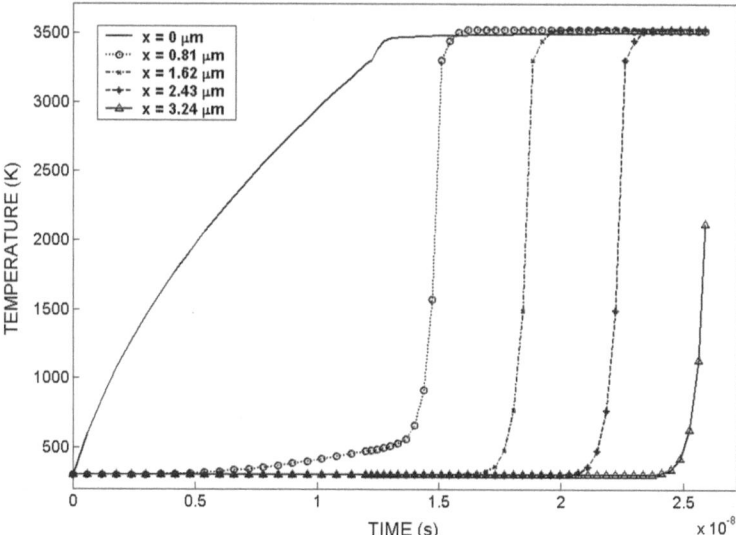

Fig. 2.11 Temporal variation of temperature at different depths

Figure 2.11 shows temporal variation of temperature at different locations inside the substrate material. In the diffusional energy transport process (for temperatures less than the evaporation temperature), temperature rises parabolically, provided that in the early heating period the rate of temperature rise is high due to internal energy gain of the substrate material. As the heating period progresses, the rate of temperature rise becomes almost steady due to diffusional energy transport from high temperature to low temperature regions. In the case of evaporation initiation, temperature remains high at interface and in the vapor side away from the interface, temperature decays rapidly. Consequently, in the vapor as well as solid regions away from the interface, the decay of temperature occurs at a high rate.

Figure 2.12 shows the temporal variation of location of moving front in reference to the initial solid surface before evaporation takes place. In the early period of evaporation, the movement of the location of the interface takes place at a high rate. As the time progresses, it becomes almost steady. This can also be observed from Fig. 2.13, in which recession velocity of the interface is shown with time. In the early period of evaporation, surface recesses at a higher rate and as the evaporation progresses, it reduces reaching almost steady value. Consequently, transient effect of evaporation process appears in the early evaporation process, as the heating duration progresses, solid surface evaporates almost at a constant rate. Since the power intensity selected is high (in order to avoid melting affect), the magnitude of recession velocity is high.

Evaporation at the Surface

The findings are presented in the light of the previous study [15]. Figure 2.14 shows cavity shape at t = 19.7 ns of the heating pulse. The presence of mushy zone particularly at the vapor–liquid interface is evident; however the size of mushy zone at solid-interface is considerably small. This is due to the latent heat of evaporation, which is significantly larger than the latent heat of melting, i.e. energy stored in the vapor–liquid mushy zone at evaporation temperature is larger than that at melting temperature.

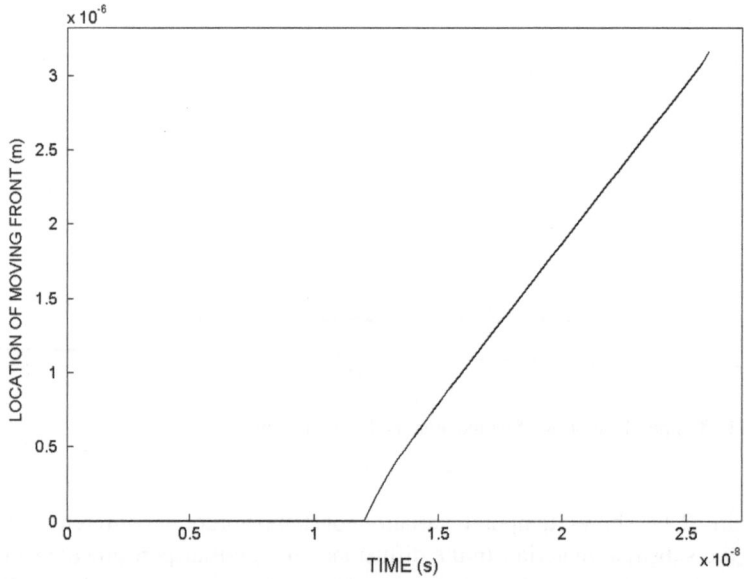

Fig. 2.12 Location of moving front with time

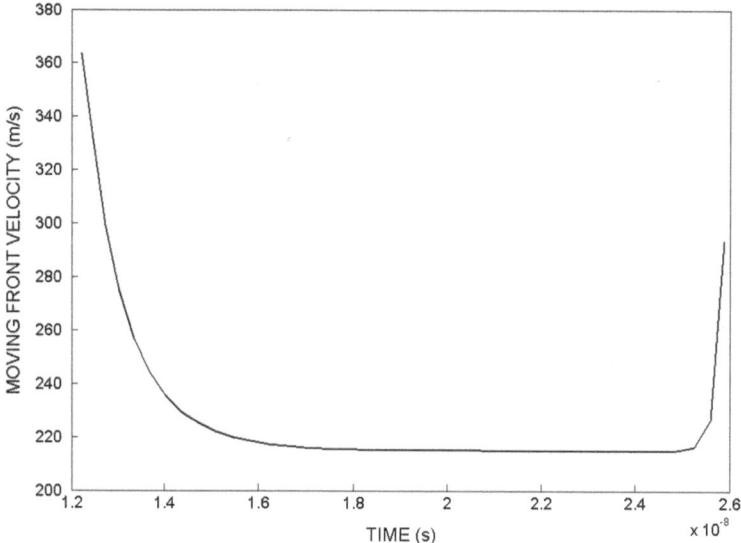

Fig. 2.13 Moving front velocity with time

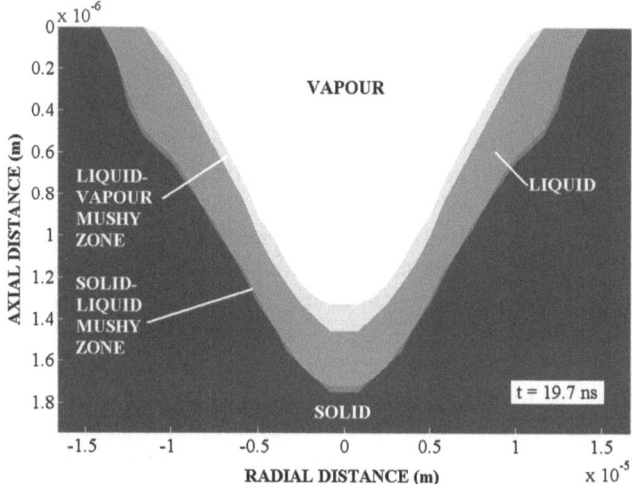

Fig. 2.14 Cavity shape and phases around the cavity for 19.7 ns of heating duration

Figure 2.15 shows contours of velocity magnitude in the cavity for different heating periods. Velocity close to the cavity symmetry axis attains high values due to high rate of mass removal rate along the symmetry axis. It should be noted that recession velocity of the cavity surface along the symmetry axis is higher than of at other radial locations. Some variation in the velocity magnitude at radial location of 1 μm along the axial distance is due to rapid expansion of vapor front close

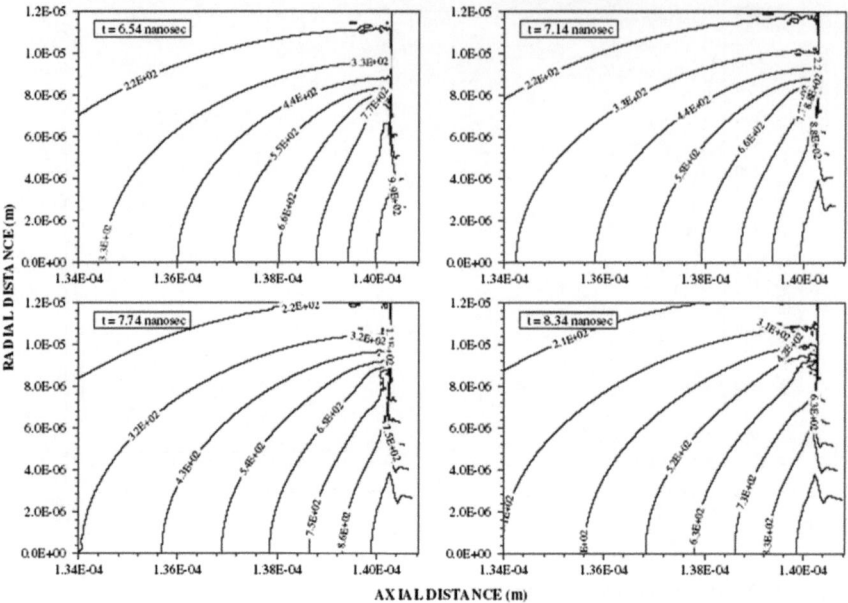

Fig. 2.15 Velocity contours in the region close to the cavity for different heating periods

to the recessing cavity surface while jet front is expanding into stagnant water. This situation results in complex flow structure in this region, i.e. suppression of jet expansion towards the cavity exit by the stagnant ambient fluid, which is water, and recessing cavity surface towards the solid bulk gives rise to high pressure vapor to expand radially. This results in pressure rise in the region close to the cavity wall. This situation can also be observed from Fig. 2.16. The pressure levels in the order of 10 GPa is generated in the cavity. The occurrence of maximum pressure close to the cavity edge is also observed from Fig. 2.16 in which pressure variation along the axial distance at symmetry axis is shown. Increasing heating duration lowers the recoil pressure in the cavity. This may occur because of one or all of the followings: (i) vapor jet penetration into the stagnant water ambient lowers the axial as well as radial momentum of the jet as the time progresses, and (ii) the cavity size increases, because of the evaporation of the cavity wall due to absorption of the laser beam, and expansion of the jet in the radial as well as axial directions enhances in the cavity lowering the recoil pressure.

Thermal Stress Analysis in the Cavity in relation to Drilling:

Thermal stress analysis is presented in the light of the previous study [3]. Figure 2.17 shows the dimensionless temperature distribution $(T^* = T \frac{k\delta}{I_0})$ along the dimensionless distance $(x^* \delta)$ inside the substrate material for three dimensionless heating periods $(t^* = \alpha \delta^2 t)$. Temperature drops sharply with increasing distance from the surface. This is more pronounced as the heating period progresses $(t^* = 29.11)$. In the early heating period, the energy gain by the substrate material

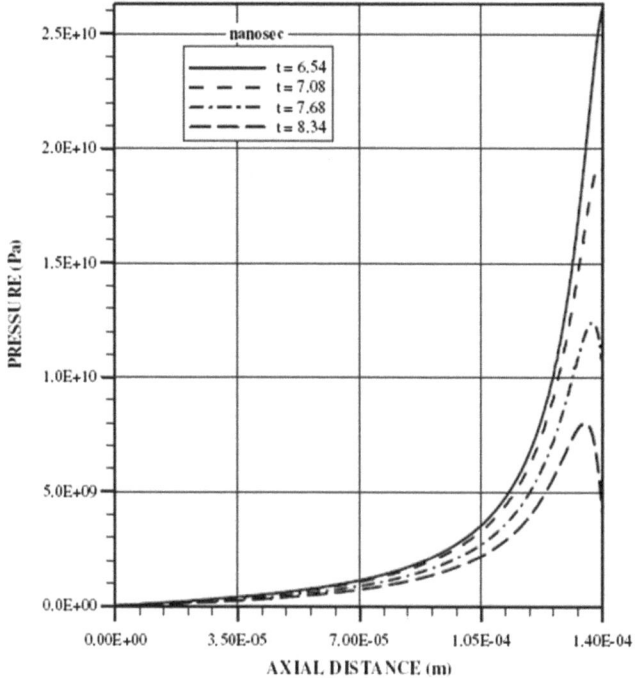

Fig. 2.16 Pressure distribution along the symmetry axis at different heating durations

Fig. 2.17 Dimensionless temperature variation along the dimensionless axial distance for different dimensionless heating periods

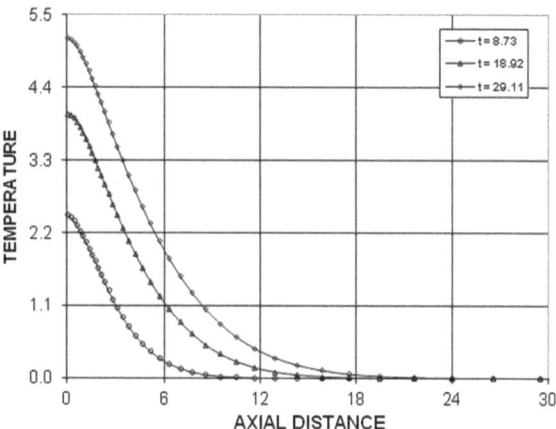

results in temperature rise in the surface vicinity. However, the temperature gradient is not considerably high to accelerate the heat diffusion from surface region to bulk of the substrate material. Consequently, in the early heating period internal energy gain of the substrate dominates the heat diffusion, which takes place from surface region to bulk of the substrate material. As the heating progress the

Fig. 2.18 Dimensionless temperature variation along the dimensionless radial distance for different dimensionless heating periods

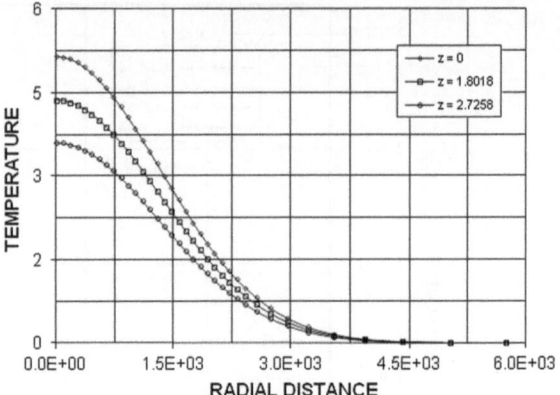

internal energy gain of the substrate material becomes high in the surface vicinity due to the absorption of irradiated energy field and temperature gradient across the surface region becomes high. The laser beam energy is absorbed within an absorption depth, which is about 1.6×10^{-8} m; therefore, for the distances extend from the absorption depth towards the solid bulk, the internal energy gain is not considerable. Moreover, the large temperature gradient in the surface region accelerates the heat diffusion from surface region to bulk of the substrate material. Consequently, as the heating period progresses the temperature gradient in the surface vicinity increases. The temperature gradient attains minimum value at some depth below the surface. At the point of minimum temperature gradient, the energy balance attains among the amount of absorbed energy, internal energy gain and energy diffused, i.e. the rate of internal energy gain increase becomes almost steady as the heating progresses. The location, where the energy balance occurs, moves away from the surface as the heating progresses. This indicates that as the heating progresses the temperature gradient increases towards the solid bulk.

Figure 2.18 shows dimensionless temperature distribution along the distance $(r^*\delta)$ in the radial direction at different dimensionless z-axis locations $(z^*\delta)$. The temperature gradient in the radial direction attains low values in the region close to the symmetry axis. This is because of the laser power intensity distribution across the surface, which is Gaussian. In this case, in the region close to the symmetry axis (close to the irradiated spot center), the amount of laser energy deposited into the substrate is high, i.e. the internal energy gain is high. However, the radial temperature gradient in this region is low, since the temperature distribution does not decay sharply in this region. Consequently, heat diffusion in radial direction in this region is not significant. Moreover, as the distance extends further away from the symmetry axis, the temperature gradient becomes less due to less energy deposited because of a Gaussian beam profile across the surface. Therefore, heat diffusion becomes as important as the internal energy gain through absorption of irradiated energy in this region. The temperature gradient attains minimum at some distance from the symmetry axis. At the point of minimum temperature gradient, internal

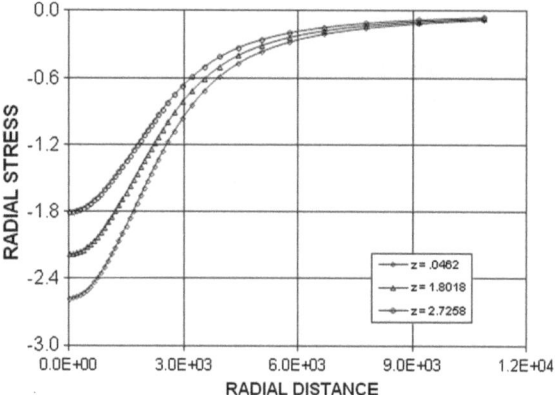

Fig. 2.19 Radial stress component in the radial direction at different z-axis locations

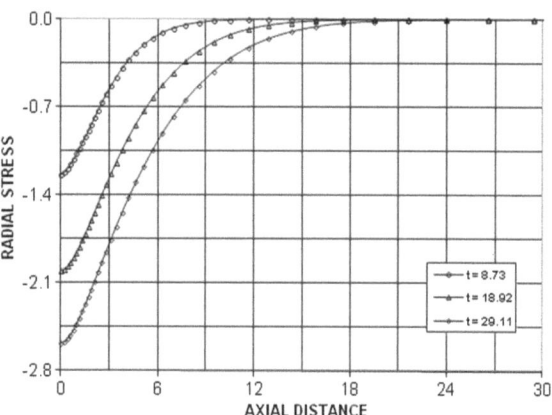

Fig. 2.20 Axial stress component in the radial direction for different periods

energy gain and conduction losses are in balance such that the rate of internal energy gain becomes steady.

Figures 2.19, 2.20 show dimensionless radial and axial stress $\sigma* = \sigma[\frac{1-\nu}{\alpha_T E}][\frac{k\delta}{I_o(1-r_f)}]$ distributions along the dimensionless distance along radial and axial directions and with dimensionless time respectively. The radial stress component is compressive inside the substrate material and it increases as the heating duration progresses. This is because of the thermal strain, which is developed during the heating period, i.e. material expands along the z-axis, since the free surface is located at $z^* = 0$ for all radial locations. Consequently, material expansion generates compression in the radial direction. The stress component is zero at the surface ($z^* = 0$), which is employed as the boundary condition in the stress analysis. The gradient of the stress component attains low values in the surface vicinity. This is because of the temperature distribution in the radial direction, i.e., temperature has low gradient in the surface vicinity of the substrate material. This is true for all heating periods. In the case of Fig. (2.7), the radial stress

Fig. 2.21 Dimensionless tangential stress component along the dimensionless distance in the radial direction at different dimensionless z-axis locations

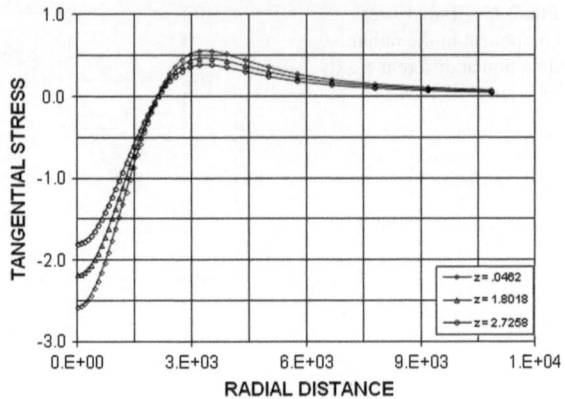

Fig. 2.22 Dimensionless tangential stress component along the dimensionless distance in the axial direction for different dimensionless times

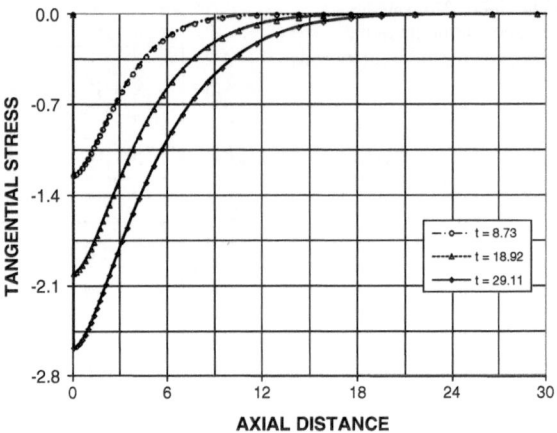

Fig. 2.23 Dimensionless equivalent stress distribution along the dimensionless radial direction at different dimensionless z-axis locations

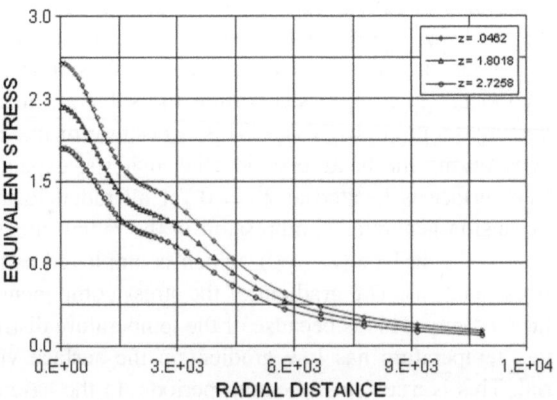

component increases with increasing heating period. The rate of increase in stress component is higher in the early heating period. In this case, the temperature rises rapidly in the early heating period due to absorption of irradiated energy, which in turn results in rapid change in temperature gradient. Therefore, the temporal variation in thermal strain becomes considerably high in the early heating period. This is more pronounced in the region irradiated by a laser beam ($z^* \leq 1$).

Figures 2.21, 2.22 how the dimensionless tangential stress component $\sigma* = \sigma[\frac{1-\nu}{\alpha_T E}][\frac{k\delta}{I_o(1-r_f)}]$ along the dimensionless distance in the radial and axial directions for different dimensionless z-axis locations and at different dimensionless heating periods, respectively. The stress component is compressive in the region close to irradiated spot center and as the radial location moves away from the irradiated spot center, it becomes tensile. The compressive behavior of the stress component close to the irradiated spot center is because of the temperature profile in the radial direction. In this case, temperature attains high values at the irradiated spot center and as the radial distance increases away from the irradiated spot center it reduces such that temperature gradient in radial direction ($\partial T^*/\partial r^*$) becomes high. This results in change of strain in this region. Consequently, positive stress levels are resulted. As the distance increases further away from the irradiated spot center stress component reduces to zero as consistent with the boundary conditions employed in the stress analysis. In the case of Fig. 2.21, the behavior of the stress component is similar to those shown in Fig. 2.20, i.e. it is compressive in the region close to the irradiated spot center while it is tensile away from this region for all the heating periods. The stress component attains high values as the heating period progresses. This is true for the compressive and tensile stress components. The stress gradient becomes higher as the heating period progresses. This is because of the development of temperature field with time, i.e. the rate of rise of temperature distribution increases in the radial direction as the heating progresses.

Figure 2.23 shows dimensionless equivalent stress distribution along the dimensionless distance in the radial direction at different dimensionless z-axis locations. The equivalent stress shows wavy appearance with radial distance. This occurs because of the influence of radial distribution of stress components. In this case, all the stress components (radial, tangential and axial components) attain high values in the region close to the surface and the gradients of the stress components along the radial direction are not the same. The stress gradient is low in the region close to the symmetry axis. This is because of the temperature distribution in the radial direction, i.e. the temperature gradient ($\partial T^*/\partial r^*$) is low in this region. Moreover, temperature gradient decays rapidly as the distance from the symmetry axis extends to $250 \leq r^* \leq 1100$, when the radial distance increases further $1100 \leq r^* \leq 3000$ the equivalent stress gradient becomes high. This because of the tangential stress component which becomes tensile in this region (Fig. (2.9). The effect of the z-axis locations on the equivalent stress is significant in the region $3000 \leq r^* \leq 9000$. In this case, equivalent stress gradient decreases at a larger rate in the surface vicinity ($z^* = 0.462$) than those corresponding to other locations in the axial direction.

Fig. 2.24 Three-dimensional
sliced view of temperature
contours after 16 s into the
cooling period. Note that the
cooling period begins after
the drilling is completed

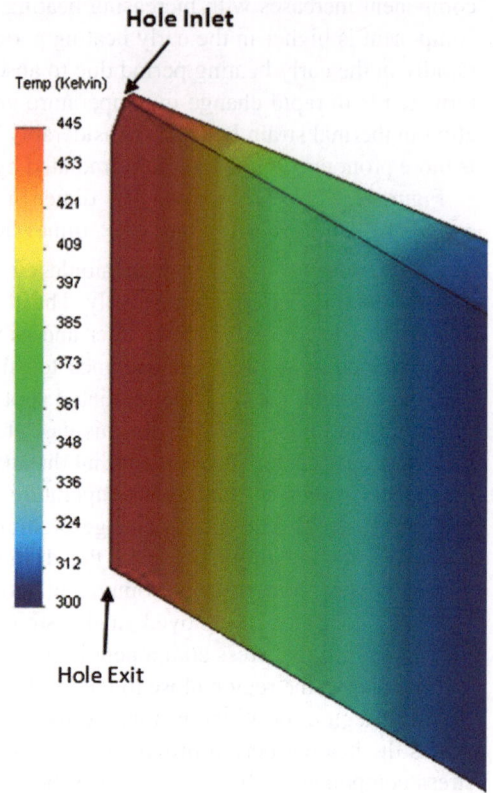

Fig. 2.25 Temperature
distribution along the
x-axis at two planes in the
workpiece

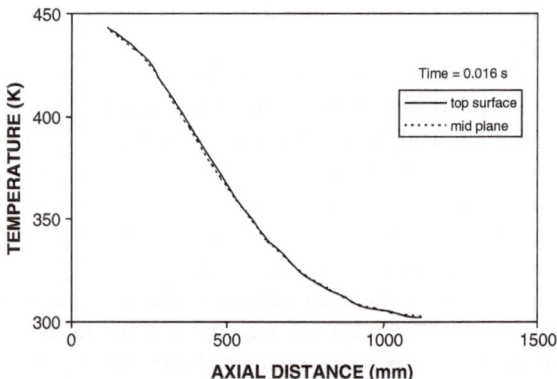

Analysis of Thermal Stress Development in the hole:

The analysis of thermal stress development is presented in the light of the previous study [4]. Figure 2.24 shows a three-dimensional view of the temperature contours around the hole perimeter, while Fig. 2.25 shows the temperature distribution along the x-axis in the region of the top and bottom surfaces of the hole during the cooling cycle. It should be noted that the cooling cycle starts once the through hole drilling is completed. Both figures show that the temperature decays gradually in the vicinity of the hole circumference, but as the distance from the hole surface increases towards the solid bulk the temperature decay becomes sharper and then decays more gradually as the distance increases further. The gradual change of temperature in the surface vicinity of the hole is associated with the heating during the hole drilling prior to the onset of the cooling cycle. In this case the internal energy gain through conduction remains high during the hole drilling. But once the hole is formed the cooling cycle initiates and heat transfer via convection from the hole surface to the hole interior remains low due to the low heat transfer coefficient. This, in turn, lowers the amount of heat transfer in this region whilst maintaining high internal energy in this region. However, the sharp decay of the temperature in the region adjacent to the hole is associated with the heat transfer from this region to the solid bulk. In this case the high temperature gradient accelerates heat diffusion from this region to the solid bulk. The attainment of low temperature decay at some distance away from the region next to the surface region is because of the low temperature gradient developed in this region; that is, thermal conduction is suppressed by the low temperature gradient in this region.

Figure 2.26 shows a three-dimensional view of the von Mises stress distribution around the hole and Fig. 2.27 shows von Mises stress distribution along the x-axis during the cooling cycle. It should be noted that the von Mises stress is shown at different y-axis locations inside the substrate material. $y = 0$ represents the free surface (top surface) and $y = 900$ µm is the bottom surface of the workpeice. The von Mises stress attains low values in the top and bottom surfaces of the workpeice. This is attributed to the expansion of free surfaces. Moreover, Fig. 2.5 shows that the thermal strain developed within the vicinity of the top and bottom surfaces results in von Mises stress variation along the x-axis in this region. It is evident from Fig. 2.5 that the von Mises stress initially reduces and then reaches the local minimum and increases to reach its local maximum. The location of the local maximum is in the region next to the hole surface, which is where the temperature gradient attains significantly high values. Consequently, the von Mises stress is high in the region where the temperature gradient is highest in the cooling cycle. Moreover, as the y-axis location along the hole wall changes and moves away from the top and bottom surfaces of the hole, the von Mises stress attains significantly high values in the surface vicinity This is because of the thermal expansion of the substrate material and the subsequent formation of strain in this region. The expansion of the top and bottom surfaces of the hole lowers the stress levels in this region; however, expansion towards the hole inner wall surface is limited, which in turn causes high strain in this region. As a result, the von Mises stress remains high in the hole surface vicinity and increasing distance along the x-axis von Mises stress reduces. The von Mises stress reduces sharply, however, in the vicinity of the surface, unlike those corresponding to the top and bottom surfaces vicinity. Moreover, as the distance along the x-axis increases further, the von Mises stress reduces gradually; therefore, strain will have developed due to thermal expansion and compression causing changes in the stress levels in the region next to the surface.

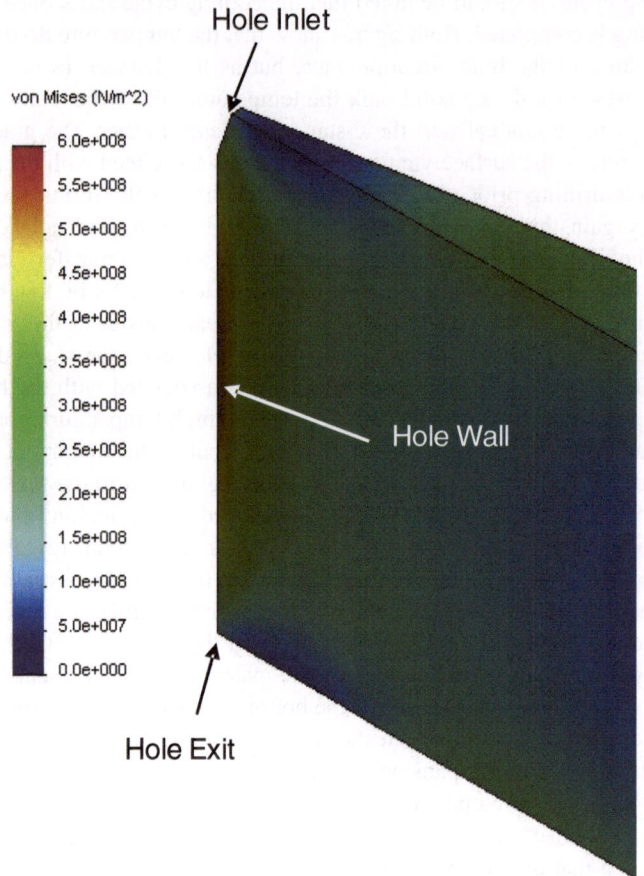

Fig. 2.26 Three-dimensional sliced view of the von Mises stress contours after 16 s into the cooling period. Note that the cooling period begins after the drilling is completed

Fig. 2.27 Temperature distribution along the x-axis at different planes in the workpiece

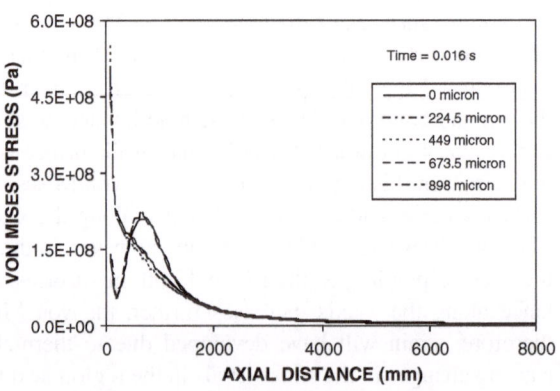

Fig. 2.28   SEM micrographs of the Nd:YAG laser drilled holes

Fig. 2.29 Optical
micrograph of the cross-
section through the Nd:YAG
laser drilled hole

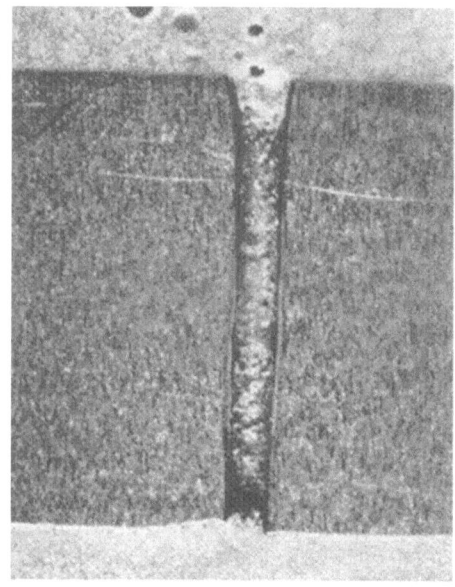

Figure 2.28 shows SEM micrographs of the top and bottom surfaces of a micro-
hole Nd:YAG laser drilled in a steel substrate, while Fig. 2.29 shows an optical
micrograph of the hole cross-section. Some surface debris around the hole circumference is clearly evident. This is because of the material ejection by the drag
forces from the hole wall vicinity. However, the amount of surface debris is considerably less than large holes drilled by millisecond laser pulses. The close examination of the micro-hole cross-section reveals that inlet and exit cones are formed
after the drilling process. The formation of the cones is due to pressure build up
in the cavity during the initial and final stages of the hole formation. In this case,
ejection of high pressure vapor and liquid material is responsible for the thermal
erosion at hole inlet and exit during the hole formation.

Fig. 2.30 First and second
law efficiencies with cutting
speed for different laser
power levels

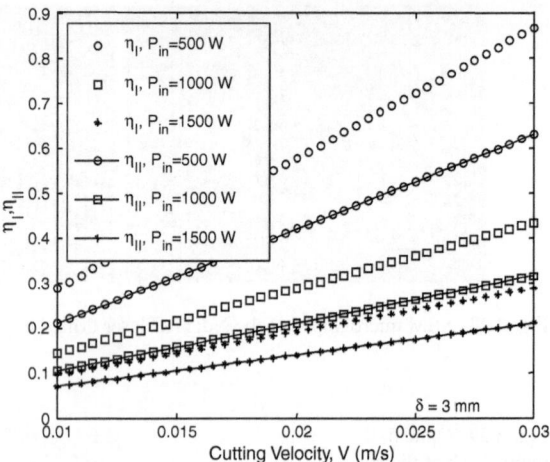

Thermal Efficiency of hole cutting:

Thermal efficiency analysis of laser hole cutting is presented in the light of the previous study [5]. Figure 2.30 shows the first and the second law efficiencies of the laser cutting process for laser cutting speeds for 3 mm thick Kevlar plate. The first law efficiency (energy efficiency) attains higher values than that of the second law efficiency. This is because of the energy available becomes less during the cutting process due to entropy generation. Since the useful work done is limited with the amount of material removed while excluding the losses associated with the heat transfer from the cutting section. It should be noted that temperature at the cutting section remains high while the region away from the cutting section in plate is at low temperature. This results in heat transfer from the cutting section to its environment. Consequently, entropy generation lowers the useful energy for the cutting process. This situation is true for all the laser power levels used in the simulations. Moreover, increasing cutting speed enhances both the first and second law efficiencies. This is because of the fact that increasing cutting speed requires increased laser power (Fig. 2.2), which improves the first law efficiency. In addition, temperatures on the cutting section remains the same at melting temperature of the substrate material for all cutting speeds, entropy generation remains the same for all cutting speeds. Since the power available increases while entropy generation rate remains the same with increasing cutting speed, the second law efficiency increases with increasing cutting speed. However, increase in the first law efficiency with cutting speed is larger than the second law efficiency. This is true for all the power levels adopted in the simulations.

Figure 2.31 shows the first and second law efficiencies with laser power for different cutting speeds for 3 mm thick Kevlar plate. The first law efficiency is higher than the second law efficiency because of the entropy generation during the cutting process. The first and second law efficiencies reduce with increasing laser power, which is more pronounced at low laser power levels. The decay of the first law

Fig. 2.31 First and second law efficiencies with laser power for different cutting speeds

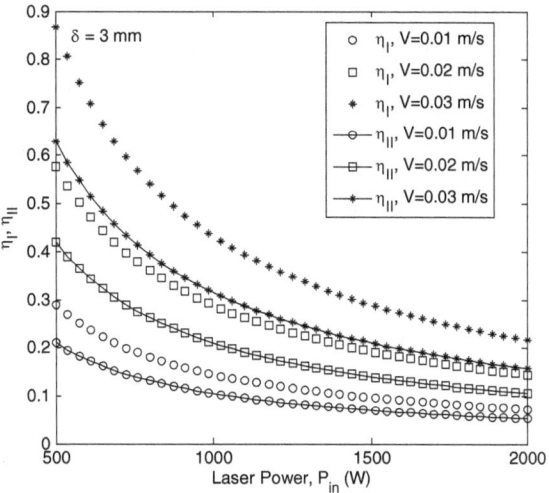

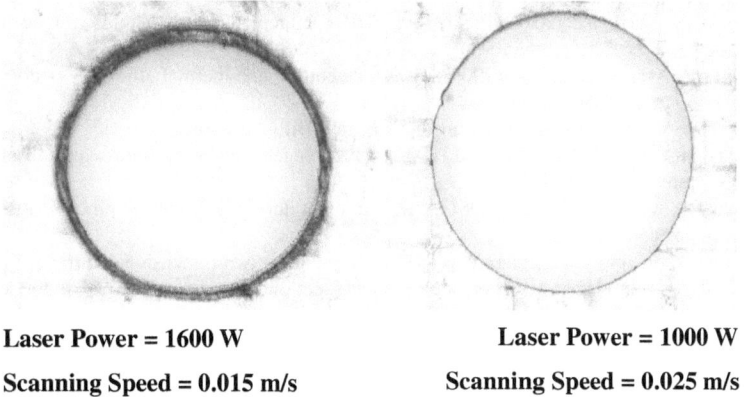

Laser Power = 1600 W

Scanning Speed = 0.015 m/s

Laser Power = 1000 W

Scanning Speed = 0.025 m/s

Fig. 2.32 Laser cut holes into 9 mm thick Kevlar plates

efficiency is associated with the power required and power input by an external laser source for the cutting process. As the power required for cutting is in the order of power input by a laser source, the first law efficiency attains high values. Increasing laser power input more than the required lowers the first law efficiency. In the case of the second law efficiency, similar arguments can be applied provided that entropy generation lowers the useful energy in the cutting section. This situation is more pronounced at low cutting speeds. Consequently, combination of low cutting speed and high laser power input lowers both the first and second law efficiencies.

Figure 2.32 shows optical photographs of laser cut holes in Kevlar plate at different laser input power and cutting speeds. It can be observed that increasing laser power input results in sideways burning around the hole circumference while lowering the cut quality. This situation is more pronounced at low cutting speeds. Consequently, as the first and second law efficiencies reduce, the resulting cut quality also reduces. In this case, sideways burning and cutting irregularities become visible around the hole circumference. On the other hand, lowering the laser power while increasing the cutting speed improves the end product quality of the cut holes, i.e., sideways burnings and cutting irregularities disappear around the cut edges. It should be noted that at high temperature combustion occurring at high laser power levels and low cutting speeds is responsible for the sideways burning.

References

1. Yilbas BS, Mansour SB (2007) Laser heating: jet emanating from laser induced cavity. Int J Therm Sci 46(4):385–398
2. Shuja SZ, Yilbas BS, Khan S (2009) Jet impingement onto a tapered hole: influence of jet velocity and hole wall velocities on heat transfer and skin friction. Int J Numer Methods Fluids 60:972–991
3. Yilbas BS, Naqvi I (2006) Laser heating and thermal stresses time exponentially heating puls case. Trans Can Soc Mech Eng 30(1):113–142
4. Yilbas BS, Shuja SZ (2010) Predictions of temperature and stress fields around the hole perimeter of laser drilled micro-holes. Lasers Eng 20(3–4):129–142
5. Sahin AZ, Ayar T, Yilbas BS (2009) Laser hole cutting and thermal efficiency analysis. Int J Exergy 6(4):592–604
6. Patankar SV (1980) Numer heat transfer. McGraw-Hill, New York
7. Yilbas BS, Shuja SZ, Budair MO (2003) Jet impingement onto a hole with constant wall temperature. Numer Heat Transf 43:843–865
8. Yilbas BS, Abdul Aleem BJ (2006) Dross formation during laser cutting process. J Phys Part D Appl Phys 39:1451–1461
9. Elkaim D, Reggio M, Camarero R (1992) Simulating two-dimensional turbulent flow by using the k-ε model and the vorticity-stream function formulation. Int J Numer Methods Fluids 14:961–980
10. Timenshenko SP, Goodier JN (1984) Theory of elasticity, 3rd ed. McGraw-Hill Book Company, Singapore, pp 476–484
11. Paek U, Gagliano FP (1972) Thermal analysis of laser drilling—IEEE J. Quantum Electron 8:112–119
12. Kovalenko AD (1969) Thermoelasticity, basic theory and applications, academy of sciences of the Ukranian USSR, Institute of Mechanics. In: Macvean DB, Alblas JB (eds) Wolters-Noordhoff Publishing Groningen, The Netherlands, pp 188–195
13. www.solidworks.com/. Accessed 8 Jan 2010
14. Al-Sulaiman F, Yilbas BS, Karatas FC, Ahsan M, Mokheimer EMA (2007) Laser hole cutting in Kevlar: modeling and quality assessment. Int J Adv Manuf Technol 38(11–12):1125–1136
15. Yilbas BS, Mansoor SB, Arif AFM (2009) Laser shock processing: modeling of evaporation and pressure field developed in the laser produced cavity. J Adv Manuf Technol 42:250–262

Chapter 3
Quality Assessment of Drilled Holes

Inconel alloys are widely used in gas power industry as materials for hot gas path components. In order to achieve a long life operation of hot gas parts, substantial cooling of the parts is essential in gas turbine engines. One of the mechanisms to make the passages for coolants in hot path components is the drilling through holes in the parts by a laser beam. In order to improve the practical laser hole drilling, an understanding of the physical process involved in drilling is essential. Considerable research studies were carried out to investigate the laser hole drilling. When modeling the laser drilling, the main emphasis was given to the laser heating process and the material ejection process during the laser-workpiece interaction. The heating mechanisms in relation to the drilling were examined extensively in the past. Some of these studies include the modeling of material response to the laser pulse [1–3]. Moreover, a number of investigators pointed out that typically 90 % of material ejected from the workpiece was in liquid state when the laser power intensity was on the order of 10^{11} W/m^2 [4–6]. Previously, information on the ejected material was recorded photographically of material spatter on slides placed between the irradiated target and the laser beam focusing lens [7, 8]. The measurements provided quantitative description of the material removal during the drilling process; however, it lacks the qualitative assessment of the drilling. Moreover, the drilling process is, in general, complicated in nature and requires further investigations.

In industrial application of laser drilling, the quality of drilled holes is the important factor. Laser drilled hole quality can be judged by internal form and taper and related geometrical features, as well as extend of heat affected zone. Consequently, the levels of the drilling parameters resulting in holes with less taper and parallel-sided walls are needed to be identified. The laser and workpiece material parameters affecting the laser hole drilling include laser output energy, pulse length, focus setting of focusing lens, drilling ambient pressure and workpiece thickness. The optimum performance in laser drilling depends on the proper selection of these factors. However, with many variables and incomplete information of the relationship between them, a statistical method is fruitful to design an

B. S. Yilbas, *Laser Drilling*, SpringerBriefs in Manufacturing and Surface Engineering, 51
DOI: 10.1007/978-3-642-34982-9_3, © The Author(s) 2013

experiment and analyze the results accordingly. Moreover, factorial experimental design methods are more secure basis and make the identification of controlling factors and interactions possible with reasonable accuracy.

3.1 Evaluation of Geometric Features of Holes

A schematic view of a typical laser drilled hole is shown in Fig. (3.1).

The hole geometry was described on the basis of seven features and judged overall by assigning to each feature a mark scored out of weighted maxima, and summing these scores. The features of the hole geometry and the marking scheme are listed below:

Resolidified Material: Marks are assigned out of 10 according to the fraction of resolidified material to hole size, i.e. a score of 10 represents no solidification around the hole and 1 corresponds to resolidified material almost block the hole.

Taper: The taper angle, as shown in Fig. (3.1), is measured without considering the inlet and exit cone angles. A mark of 0 indicates taper of 90° and 4 corresponds to a parallel-sided hole.

Barreling: The marks are assigned in between 0 and 4. A mark of 4 represents a straight hole while 0 indicates an even hole.

Inlet Cone Angle: It is measured as ratio of cone angle to 180°. A mark of 1 indicates no inlet cone and 0 corresponds to a cone angle of 180°.

Exit Cone Angle: In general, the exit cone is small in size and difficult to measure with accuracy. Therefore, marks are given through visual inspection and mark of 1 indicates no exit cone while 0 represents a large cone angle.

Surface Debris: This is assessed according to the amount of material resolidified at the surface of the workpiece near to the hole circumference. A mark of 1

Fig. 3.1 Features of laser drilled holes

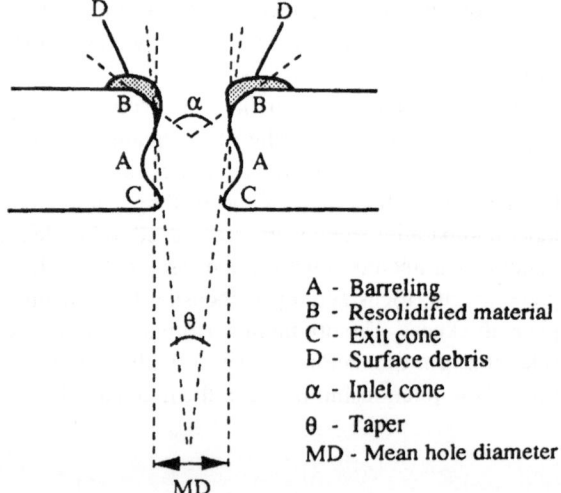

A - Barreling
B - Resolidified material
C - Exit cone
D - Surface debris
α - Inlet cone
θ - Taper
MD - Mean hole diameter

is given with no or very small amount of surface debris and mark of 0 is assigned with large amount of surface debris.

Mean Hole Diameter: The mean hole diameter is measured under the microscope and the maximum value of mean hole diameter among all holes is considered and a mark of 3 is assigned. The minimum hole diameter is given a mark of 1.

Overall Quality: The overall quality of the drilled hole is obtained summing the all marks assigned each hole feature, i.e.:

$$q = Re + Ta + Ba + Ec + Ic + Sd$$

where q, Re, Ta, Ba, Ec, Ic, and Sd are marks of overall quality, resolidification, taper, barreling, exit cone, inlet cone and surface debris, respectively. Moreover, for blank holes a mark of 3 is subtracted from the overall quality.

3.2 Factorial Analysis in Relation to Laser Drilling

The qualitative and quantitative analyses can be carried out to assess the effect of parameters on the quality of the drilled hole. Although many factors affecting the hole quality, only four factors at four levels are considered as an example, which are given in Table 3.1.

3.2.1 Qualitative Analysis

The qualitative analysis is involved with the qualification of the features of the hole geometry that cannot be measured concretely; but, a numerical scheme can be incorporated to distinguish the effects of the parameters on the hole features. The mathematical expression representing the factorial design can be expressed as:

$$X_{i,j,k,l} = P_i - F_j - E_k - T_l - P_iF_j - P_iE_k - P_iT_l - F_j$$
$$E_k - F_jT_l - E_kT_l - P_iE_kF_j - P_iE_kT_l - F_jE_kT_l - \varepsilon_{ijkl} \quad (3.1)$$

where X_{ijkl} is the response (value given to the hole features), P_i, F_j, E_k, T_k are the main effects of pressure, focus settings, laser output energy and workpiece thickness, P_iF_j, P_jE_k, P_iT_l, F_jE_k, F_jT_l, E_kT_l, $P_iE_kF_j$, $P_iE_kT_l$, $F_jE_kT_l$ are the first and second order interactions of selected factors, and ε_{ijkl} is the random error in the

Table 3.1 Parameters and their levels employed in the experiment

Pressure (mm Hg)	Focus setting (mm)	Laser output energy (J)	Thickness (mm)
150	50	15	0.6
250	50.5	17	0.8
350	51	19	1.0
760	51.5	21	1.2

experiment provided that i, j, k, l represent the levels of pressure, focus settings, laser output energy, and thickness respectively. It should be noted that:

$$P_i \times F_j = F_j \times P_i$$
$$P_i \times F_j \times E_k = F_j \times P_i \times E_k$$

This is valid for all the other interactions.

To calculate the main effects of each factor and the effects of interaction, it is necessary to determine the mean corrected sum of squares of response at all levels of factors. However, the mathematical arrangements of qualitative analysis are not given here, but it is referred to [9].

The significance of the effect of factors and interactions is tested by making the initial hypothesis that the mean squares of the effects or interactions are from the same populations as the error mean square, which is that all the mean squares are on a comparable basis with respect to error variance. It should be noted that the mean squares of effects and interactions are linear functions of observation. Since there are no repeats, the remainder mean square are representing the higher order interactions being used as an estimate of the error variance to test the significance of the effects calculated by means of the F-test [9]. The mean square of the effects is determined by dividing the corrected sun of squares by an appropriate number of degrees of freedom. In this case the variance ratio is:

$$F_{r,s} = \frac{\text{Mean sum of squares}}{\text{Error mean sum of squares}}$$

where r and s are the degrees of freedom of the mean sum of squares and the error mean sum of squares, respectively.

If the calculated variance ratio of any effect exists in the value in the F-test table [9] at corresponding r and s degrees of freedom, the effect is significant at that level.

3.2.2 Quantitative Analysis

A quantitative factor is a factor for which the levels may be represented in terms of continuous points. The effect of such a factor is a mathematical function of the levels of the factor. This can be written in a polynomial form [9] as:

$$X = a_0 y + a_1 y^2 + a_2 y^3 - \cdots\cdots\cdots + a_n y^{n+1} \tag{3.2}$$

Table 3.2 Orthogonal coefficients of polynomial equations

	Y_1	Y_2	Y_3	Y_4	η	γ
Linear	-3	-1	1	3	20	2
Quad.	1	-1	-1	1	4	1
Cubic	-1	3	-3	1	20	10/3

where X is the response, y is the level of factor, and n is the number of levels selected. The coefficients in the polynomial equation are estimated by the least square technique. Since four levels were selected, the values of the orthogonal coefficients are given Table 3.2 where η and γ are the sum of squares of the coefficients and a scale factor, respectively. The actual values of the constants in the orthogonal form of the polynomial are derived from $\Sigma\beta'X$, $\Sigma\gamma'X$, $\Sigma\delta'X$ multiplying the corresponding quantity of η/γ [9]. However, during the calculations of the variances, it is convenient to drop the multiplier γ/η, which is needed only to be introduced when the actual numerical values of the constants are required.

The details of the mathematical arrangements of quantitative analysis are given in [9]. The variance ratio is determined similar to that obtained for the quantitative analysis, i.e.:

$$F_{r,s} = \frac{\text{Mean sum of squares}}{\text{Error mean sum of squares}}$$

A computer program is developed to calculate the variance ratios due to qualitative and quantitative analyses.

3.3 Experimental

In order to introduce the factorial analysis for assessing the laser drilled hole geometry, an experiment is carried out and Inconel 617 alloy sheet is drilled. An Nd: YAG laser delivering energies in between 10 and 21 J within 1.48 ms pulses was used to irradiate the workpiece while nominal focal length of 51 mm lens was used to focus the laser beam on to the workpiece surface. A graduated barrel adjusted a focus setting of the lens. The knife-edge and optical methods were used to determine the beam waist position reference to the lens holder. The details of the measurement are given in [10]. The mode of laser beam was TEM00 and small variation in temporal distribution of the laser power intensity was observed after the each pulses. A vacuum cell was designed and was build to locate the workpiece and regulate the assisting ambient gas pressure during drilling. Air at sub-atmospheric pressures was used as drilling ambient. The holes drilled in Inconel 617 alloy were cross-sectioned, mounted in a perspex, and polished before examining under the microscope. The evaluation of hole geometric features was achieved using optical microscopy and scanning electron microscopy (SEM).

3.4 Findings and Discussion

Laser hole drilling in Inconel 617 alloy is carried out. An F-test was introduced to the results obtained from the factorial analysis to identify the significant levels of the main parameters and their interactions on the geometric features of laser drilled holes. The terms most significant, very significant, and significant refer to the minimum correlation levels of 0.99, 0.98 to 0.95, and 0.94 to 0.90 of the factors, respectively. Moreover, in the qualitative analysis "Main" effects are zero order effects (independent of each other). The first order interaction demonstrates that the effects of the interacted factors on observation are dependent on each other. The quantitative and qualitative analyses results are given under the appropriate headings.

3.4.1 Qualitative Analysis

Table 3.3 gives the F-test results for the parameters pressure (P), focus settings (F), laser output energy (E), and workpiece thickness (T). The discussion for the test results are given under the headings relevant to the geometric features of the hole drilled.

Resolidified Material: The main effect of pressure is significant, the first order interactions of pressure-thickness is most significant, pressure-focus setting and pressure-thickness are significant, and second order interaction of pressure-focus setting-energy is significant. This indicates that the ambient pressure has a significant effect on the resolidified material. In this case, reducing ambient pressure generates a pressure differential across the evaporated front and its surroundings. Consequently, the material removal rate increases from the cavity in the initial stage of the hole formation. Once the hole is drilled in the workpiece, the pressure in the cavity drops considerably. Despite the low pressure in the cavity, the low ambient pressure results in molten metal to drag towards the hole inlet. The molten metal, then, solidifies as soon as the laser pulse ceases. Therefore, the

Table 3.3 Qualitative analysis results (> represents the parameter is not significant)

	Resolid. material	Taper	Barrel.	Inlet cone angle	Exit cone angle	Surface debris	Mean hole diam.	Overall quality
P	0.9	0.99	0.9	0.9	>0.9	>0.90	0.99	0.90
F	>0.9	0.99	0.99	0.99	0.99	>0.90	0.99	0.99
E	>0.90	0.90	0.90	0.90	>0.90	>0.90	0.99	0.90
T	>0.90	0.99	0.99	0.99	0.99	>0.90	0.99	0.99
PF	>0.90	0.99	0.90	>0.90	0.90	0.90	>0.90	0.95
PE	>0.90	>0.90	>0.90	0.90	>0.90	>0.90	>0.90	>0.90
PT	0.99	0.99	>0.90	0.99	>0.90	>0.90	0.99	0.95
FE	>0.90	>0.90	0.95	0.95	>0.90	>0.90	>0.90	0.90
FT	>0.90	0.95	>0.90	0.99	0.99	>0.90	0.99	0.95
ET	>0.90	0.95	>0.90	0.95	0.99	>0.90	>0.90	0.90
PFE	>0.90	>0.90	>0.90	>0.90	>0.90	>0.90	>0.90	>0.90
PFT	0.90	0.90	<0.90	0.90	0.99	>0.90	0.95	0.95
PET	>0.90	>0.90	>0.90	>0.90	>0.90	>0.90	>0.90	>0.90
FET	>0.90	0.95	>0.90	>0.90	>0.90	>0.90	>0.90	>0.90

molten metal forms the resolidification at the hole exit. The first order interaction between the pressure-thickness has most significant effect on the resolidified material. This shows that resolidification around the hole exit increases considerably once the thickness increases. This is due to that the amount of molten metal in the cavity, which becomes more as the thickness increases. Therefore, the amount of molten metal, which is dragged towards the hole exit, increases. This process is also effected by laser output energy, since the second order interaction of pressure-focus setting-laser output energy is significant.

Taper: The main effects of all the parameters are found to be most significant for taper, except energy, which has only significant effect on taper. Almost all the first order interactions have most significant effect on taper while the second order interaction of pressure-focus setting-thickness has significant effect on taper. Therefore, taper increases as ambient pressure reduces, in which case the rate of mass removal from the cavity increases. In addition, the focus setting and thickness have similar effect on the taper. In this case, energy available at the workpiece surface varies as the focus setting changes. This influences the energy coupling at the workpiece surface. Hence, it is the energy coupling mechanism, which influences taper. Increasing thickness generates long cavity length before the full penetration of the workpiece is achieved. In this case, the pressure generated due to superheating of liquid becomes considerably high, which in turn accelerates the mass removal from the cavity and enhances the liquid surface propagation into the workpiece. Consequently, high rate of material removal from the workpiece results in large taper. The first order interaction indicates that the coupling effects of pressure and thickness, pressure and focus setting, and thickness and focus setting are significant. Therefore, the formation of taper depends on the individual parameters as well as their interactions.

Inlet Cone Angle: Focus setting and thickness are found to be most significant parameters while pressure and energy are significant parameters in the case of inlet cone angle. Inlet cone angle is, in general, related to focus setting since the larger beam waist at the workpiece surface generates large inlet cone angles. The pressure inside the cavity before the through hole formation is also important. The pressure in the cavity increases as thickness of the workpiece increases. In addition, the differential pressure across the evaporating front and its surroundings increases due to sub-atmospheric ambient conditions, i.e. the mass removed from the cavity erodes thermally the exiting section of the cavity, which in turn results in large inlet cone angle. Therefore, the coupling effect of pressure and focus setting influences the inlet cone angle formation considerably.

Barreling: The focus setting and thickness have most significant effect on the barreling. The main effects of pressure and energy are found to be significant. Barreling defines the how parallel-sided the hole drilled. Thickness and focus setting contributions are more than pressure and energy contributions for barreling formation. This suggests that formation of parallel-sided hole is related to the pressure rise inside the cavity before through hole formation rather than the mass removal rate, i.e. the regulated mass removal rate results in improved barreling than the fast rate of mass removal. The first order interaction between pressure and focus setting shows that there exists a coupling effect of these parameters on barreling as the combinations of focus settings and pressure vary.

Exit Cone: The focus setting and thickness have the most significant effect while pressure and energy have significant effect on the exit cone. The first order interactions of focus setting-thickness, and focus setting–energy are the most significant. This finding indicates that thickness, focus setting, and energy have significant contribution to the exit cone formation. In this case, the thick workpiece results in large pressure development in the cavity during the hole formation. This is effected by the beam waist position and laser output energy such that large exit cone is resulted for thick workpieces. In addition, the exit cone becomes small for some combinations of these parameters.

Surface Debris: Surface debris is a measure of amount of material at the surface around the hole circumference. In almost all cases, the surface debris is not observed. This is because the low ambient pressure. In this case, the material leaving the hole inlet and exit is accelerated due to the pressure differential. This results in less or no debris at the workpiece surface. The first order interaction of pressure and focus setting has a significant effect on the surface debris.

Mean Hole diameter: The main effects of all the parameters are found to be most significant. The effects of first order interactions of pressure-thickness, and focus setting-thickness are most significant. The size of the mean hole diameter depends on the amount and the rate of mass removal from the cavity during the hole formation. In this case, increase in pressure in the cavity and low ambient pressure enhance the amount of mass removed and its rate. In addition, the energy coupling at the workpiece surface alters the rate of mass removal, i.e. improved energy coupling at the workpiece surface accelerates the mass removal rate. Moreover, the effects of pressure and focus setting vary with thickness, since the first order interactions of these parameters are most significant.

Overall Quality: Overall quality is effected by all parameters provided that the main effects of focus setting and thickness are the most significant while pressure and energy are significant parameters. The first order interactions of focus setting-thickness is very significant while pressure-focus setting is significant. This indicates that overall quality is not only effected by each parameter, but the combinations of pressure-focus setting, and focus setting-thickness also alter the overall quality.

3.4.2 Quantitative Analysis

The quantitative analysis gives the functional relation between the affecting parameters and the response (geometric features of the hole drilled). Table 3.4 gives the quantitative analysis results. The relation for each parameter is given below.

Pressure: The linear effect of pressure found to be most significant for mean hole diameter, very significant for barreling, and significant for inlet cone. The quadratic effect of pressure is most significant for taper, very significant for inlet cone. The cubic effect of pressure is very significant for resolidified material and it

Table 3.4 Quantitative analysis results (> represents the parameter is not significant)

		Resolid. material	Taper	Barrel.	Inlet cone angle	Exit cone angle	Surface debris	Mean hole diam.	Overall quality
P	Linear	>0.90	>0.90	0.95	0.90	>0.90	>0.90`	0.99	0.90
	Quad.	>0.90	0.99	>0.90	0.95	>0.90	>0.90	0.99	>0.90
	Cubic	0.95	>0.90	>0.90	>0.90	>0.90	>0.90	0.90	>0.90
F	Linear	>0.90	0.99	0.99	0.99	0.99	0.90	0.99	0.99
	Quad.	>0.90	>0.90	>0.90	>0.90	>0.90	>0.90	>0.90	>0.90
	Cubic	0.95	>0.90	0.90	>0.90	0.90	>0.90	>0.90	>0.90
E	Linear	>0.90	>0.90	>0.90	0.99	>0.90	>0.90	0.99	>0.90
	Quad.	>0.90	0.95	>0.90	>0.90	>0.90	>0.90	>0.90	0.90
	Cubic	>0.90	>0.90	>0.90	>0.90	>0.90	>0.90	>0.90	>0.90
T	Linear	0.95	0.99	0.99	0.99	0.90	>0.90	0.99	0.99
	Quad.	>0.90	0.99	>0.90	>0.90	>0.90	>0.90	0.99	0.90
	Cubic	>0.90	0.90	0.95	>0.90	>0.90	0.90	0.90	>0.90

is significant for mean hole diameter. This suggests that taper and inlet cone vary quadratically with pressure while barreling varies linearly with pressure. The cubic relation exists between the resolidified material and pressure.

Focus Setting: The linear effect of focus setting is very significant for all hole geometric features, except surface debris. In this case, the linear relation exists between the resolidified material, taper, barreling, inlet cone, exit cone, mean hole diameter, and overall quality with focus setting, i.e. small variation in focus setting results in small variation in features of hole geometry, since the relation between the hole features and focus setting is linear.

Laser Output Energy: The linear effect of energy is most significant for inlet cone and mean hole diameter. The cubic effect of energy is significant for taper. Consequently, as energy varies, the inlet cone and mean hole diameter change with energy at almost same rate.

Thickness: The linear effect of thickness is most significant for all the features of hole geometry, except surface debris, which varies in cubic form with thickness. Some of quadratic and cubic effects of thickness are also found to be most significant for some hole features. This suggests that although the linear effect of thickness is dominant, but as far as the interactions are concerned this effect may not be linear.

References

1. Duley WW (1983) Laser processing and analysis of materials, 1st edn. Plenum Press, New York
2. Hoadle AFH, Rappaz M, Zimmermann M (1991) Heat-flow simulation of laser remelting with experimental validation. Metall Trans B 22B:101–109
3. Yilbas BS, Sahin A, Davies R (1995) Laser heating mechanism including evaporation process initiating the laser drilling. Int J Mach Tools Manuf 35(7):1047–1062

4. Wei PS, Chian LR (1988) Molten metal flow around the base of cavity during a high energy beam penetrating process. Int J Heat Mass Transf 110:918–923
5. Yilbas BS (1998) Particle ejection during laser drilling of engineering metals. Lasers Eng 7:57–67
6. Yilbas BS (1995) Study of liquid and vapor ejection processes during laser drilling of metals. J Laser Appl 17:147–152
7. Yilbas BS, Sami M (1997) Liquid ejection and possible nucleate boiling mechanisms in relation to laser drilling process. J Phys D, J Appl Phys 30:1996–2005
8. Yilbas BS (1985) The study of laser produced plasma behavior using streak photography. Jpn J Appl Phys 24:1417–1420
9. Davies OL (1978) The design and analysis of industrial experiments, 2nd edn. Longman Group Limited, New York
10. Yilbas BS, Yilbas Z (1996) An optical method to measure the pulsed laser output power intensity distribution in the focal region, measurement. J Int Meas Confed 17(3):161–172

Chapter 4
Laser Drilling and Plasma Formation at the Surface

In laser drilling process, plasma, consisting of charged ions, electrons, and neutral atoms is formed above the surface. Drilling process is effect by the size and transient nature of the surface plasma, since the plasma partially absorbs the incident radiation and acts like a heat source contributing to the drilling process. However, formation of the excess plasma modifies the hole geometry at inlet and lowers the drilled hole quality. In order to reduce the difficulty of a theoretical investigation of laser beam plasma interactions a framework is described which resolves the interaction into zones, each of which may be considered separately. Although, in practice an alternative approach has been followed, in principle it is possible to treat the interaction as a single, if complex, event suitable for numerical solutions since the equations of hydrodynamics and heat conduction are valid through the domain of the interaction. A basis for a numerical model is to consider the material throughout the effected region to be governed by a single equation of state and the surface to be a region of rapid but continuous change in flow variables [1]. An approximate equation of state suitable for this approach may be derived from the theory of non-perfect gases. For example, the cohesive forces bonding liquid or solid phase atoms may be approximated by a Lennard-Jones type potential function. This permits a homogeneous transition of material to a vapor state to be governed in a way which allows approximately for latent heat. However, a simpler and more fruitful approach is to consider the interaction to comprise four zones, as indicated in Fig. 4.1, each of which may be examined in the light of distinguishing characteristics.

4.1 Geometric Structures of Plasma Plume

Interface Zone: This zone spans the hole penetration boundary which separates dense, effectively stationary material, from rapidly expanding vapor.

Jet Zone: This extends from distances of about 6–9 mm above the target surface to the leading edge of the expelled vapor. High speed plume is typical of free jets formed when vapor is discharged from a nozzle or orifice. These become turbulent

B. S. Yilbas, *Laser Drilling*, SpringerBriefs in Manufacturing and Surface Engineering,
DOI: 10.1007/978-3-642-34982-9_4, © The Author(s) 2013

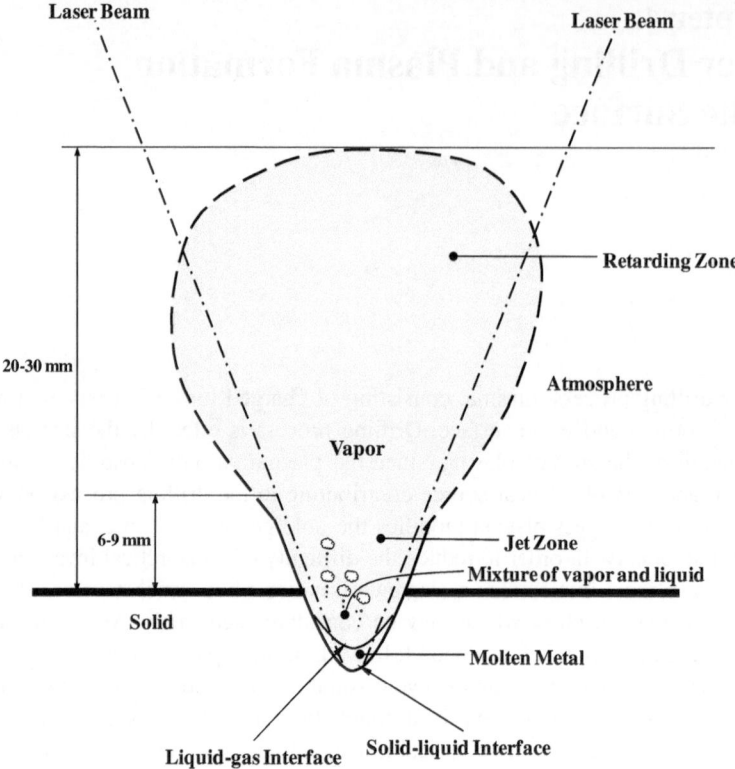

Fig. 4.1 A schematic view of laser produced plasma during the drilling

a short distance from the point of discharge and some mixing with the surrounding fluid occurs. As a result the jet spreads and some fluid from the surrounding is entrained so the mass flow in the jet increases with distance from the orifice. The increased mass flow requires a reduction of jet velocity in order to conserve the total momentum of the jet. The numerical study of jet emanating from the laser produced cavity in relation to drilling is presented in the previous study [2]. The findings revealed that the jet velocity of the order of 350 m/s occurs in the jet zone. However, using the sub-atmospheric pressure of jet ambient modifies the jet speed so that the temporal characteristics of the material ejected from the laser produced cavity change while influencing the drilling quality.

Flare Zone: This zone adjoins the interface (which may be typically 1–3 mm below the original surface plane at the end of a pulsed laser drilling interaction and extends to a maximum of 2–4 mm from the target surface.

An important observation for laser pulse drilling with millisecond pulses, which will be referred to again, is that in this region the vapor appears in highly luminous flares at intervals of 10^{-5} s, remaining luminous for a similar interval. These flares are visible in the streak photographs. Unlike vapor in the jet, the vapor near and below the surface may not be removed by a transverse jet and consequently forms an integral part of the interaction. In addition to an experimental study of the above-surface vapor, a theoretical investigation was carried out to obtain a clearer picture of events in the flare zone, using a model for the vapor/beam interaction [3].

Bulk Zone: Below surface processes have not been monitored during interactions and so observations are restricted the final condition of the workpiece. Etched crater cross-sections show that depth of liquid layer on hole walls are typically of the order of 10^{-5} m [4]. This depth provides an indication of the domain of the bulk zone in which conduction effects dominate over material motion effects.

The principal problems associated with the analysis of this zone is the transient heat conduction effects which arise when some of the energy deposited in the hole wall surface skin diffuses into the bulk, preceding the advance of the hole walls. Although the process is similar at all points of the hole boundary, the region of the interface zone is of particular interest since here the hole wall velocity and energy deposition rate have their maximum values and this combination produces large temperature changes over small distances.

4.2 Measurement of Some Plasma Characteristics in Laser Drilling

Interaction mechanism is complex and involves with complicated mathematical modeling, which may not be solved without making many assumptions and simplifications. Consequently, experimental data for the characteristics of the evaporating front are useful for improved understanding of the drilling process.

4.2.1 Laser

An Nd:YAG laser delivering pulses at 1.48 ms and energy range 15–25 J was employed to irradiate the workpiece surface. A lens of 51 mm focal length was used to focus the laser beam. The focal position of the laser beam was set to obtain the best hole drilling. Copper and nickel were used as workpiece material.

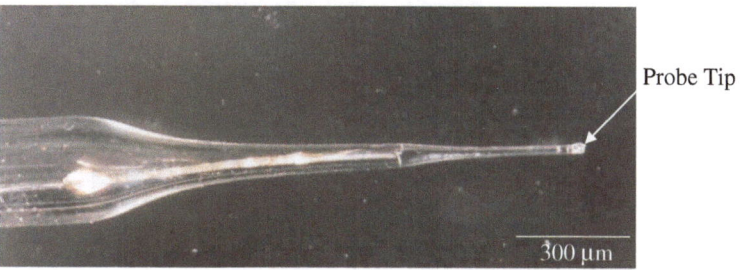

Fig. 4.2   Photograph of a Langmuir probe

4.2.2 Langmuir Probe

A single-plane Langmuir probe was used (Fig. 4.2). The probe consisted of a tungsten wire of 30 μm sealed inside a glass tube of 0.2 mm diameter at the tip. This cross section prevented the disturbances to the vapor front that may be caused by probes of large diameter. The tip of the probe was set in a position near the edge of the laser irradiated spot. The probe was arranged to move in the transverse direction. This enabled probe to monitor vapor plume size. The probe surface was frequently cleaned during the experiments in order to reduce the effect of deposits on the tip of the probe.

The Langmuir probe was connected to an electronic circuit and oscilloscope to display the ionization characteristics of the vapor front. The circuit diagram is shown in Fig. 4.3. The current–voltage characteristics by the ionization state of the vapor front were determined by varying the circuit supply voltage by 1 V increments from −12 to +6 V. The typical probe output is shown in Fig. 4.4. The details

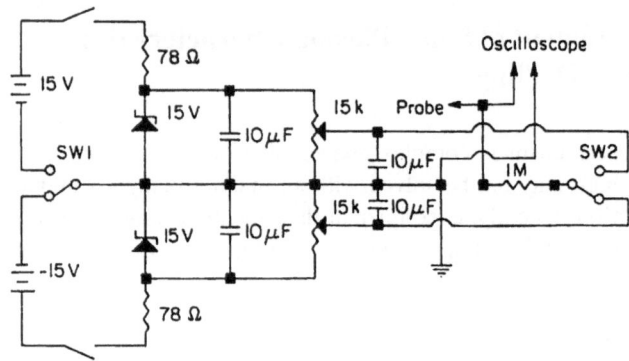

Fig. 4.3 Electronic circuit diagram for the Langmuir probe

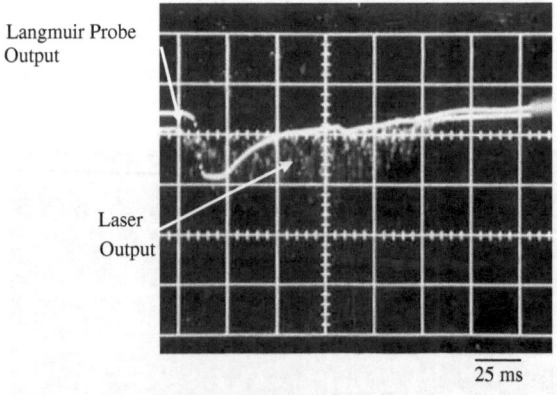

Fig. 4.4 The Langmuir probe and laser output

of the Langmuir probe measurements are given in [5]. The temporal plume height was measured by using the Langmuir probe output. Once the evaporating front reached the probe tip, then the probe gave response. Since the probe and laser source were triggered at the same time; therefore, time taken by the front to reach the Langmuir probe tip could be determined easily, i.e. it is the time shift between the laser pulse initiation and the first probe response. Similarly, the evaporating front velocity can be obtained by dividing the plume height by the time taken to travel. The velocity distribution can be obtained by moving the probe gradually in the transverse direction across the irradiated surface.

4.2.3 Streak Photography

The streak camera used was a Strobodrum streak camera and it rotated at 1.5 m strip of 35 mm film at 3000 rpm for film speed of 75 m/s. The camera was positioned so as to fill the width of the film with the event. The workpiece was on one edge of the film and the glass slide, protecting the lens from a liquid ejection, on the other edge. The flash contact on the camera shutter was used to trigger the laser. A delay circuit was designed to give up to a 20 ms delay between flash contacts closing and laser firing. The measurement of mean evaporating front velocity was carried out as follows: The film speed is 75 m/s and the scaling factor is 2.09, which gives the actual length equal to 2.09 times on the streak film. Therefore, actual distance traveled by the evaporating front is equal to multiplication of the length on the film and the scale factor. The mean evaporating front velocity can be determined as:

$$V = 156 \times \tan \theta \ (\text{m/s})$$

where θ is the angle between the free surface of the workpiece and the plane in which the front travels.

4.3 Absorption and Transmittance Characteristics of Plasma Plume

The plume has two-fold effects on the laser machining process. These are: (i) partially blocking, absorption, scattering, and deflection of the incident laser beam, and (ii) plume heating of the workpiece surface. It was argued that high density plume resulted in high degree of blocking and absorption of the reference laser beam and that the plume blocking, absorption and heating had the coupling effect on the machining process. When probing the transmittance of the surface plume, in general, He–Ne laser beam was used as a reference beam [6]. However, the surface plume is generated due to the incident laser beam, which has a different characteristic than that of the reference beam. Therefore, to obtain the realistic transmittance properties of the surface plume, probing of the reference beam of the same kind to the incident laser beam is necessary. Consequently, the present work was conducted to determine the transmittance characteristics of the incident laser beam by sampling the incident beam by a beam splitter and using it as a reference beam. To achieve this, a Nd:YAG laser was used as an incident beam to irradiate the target, which in turn produced the surface plume. The experiment was extended to include different laser pulse energy and pulse length combinations, and different workpiece materials. Consequently, the effects of pulse parameters as well as workpiece materials on surface plume transmission process were analyzed. The previous study showed that liquid ejection occurred, because probably, pressure developed in the cavity and explosions resulted from nucleation of vapor bubbles in the liquid zone [7]. In addition, it was also evident from the lens protecting glass that a region of globules imbedded in a background of finer material. Under a microscope, some globules appeared as centers of radially extending tracks, suggesting that they were caused by liquid particles, which collided with the cover glass. Consequently, in laser machining applications, the contribution of liquid vapor ejection to the total mass removal mechanism should be considered. Therefore, investigation should be extended to include the streak photograph of ejection during the laser machining process. Accordingly, an experimental program was developed which involved the use of a rotating drum camera to record streak photographs of laser-workpiece interaction process.

In order to investigate the transmittance characteristics of the incident laser beam, the plume was probed in a plane parallel to the workpiece surface and normal to the Nd:YAG laser beam axis. The transmitted beam power was monitored for a range of conditions of laser and material properties. The plume was generated during the interaction between a metal target and a focused Nd:YAG laser beam. The laser delivered the output energy in between 7–20 joules within 1.2–1.7 ms pulses. Therefore, the laser output power during each shot was of the order of 104 W. The laser beam was focussed by a Dallmeyer air spaced doublet with a nominal focal length of 51 mm and the spherical aberration was corrected to ¼ λ (λ being the wavelength) at the edge of the aperture. The $1/e$ points of the power intensity distribution correspond to 0.3 mm from the center of the irradiated spot as measured in the previous study [8]. The reference beam was sampled from the incident beam using a beam splitter, only 1 % of the incident beam was sampled as a reference beam.

Typical temporal output power intensity of incident and reference beams is shown in Fig. 4.2. However, the laser output did not provide a continuous single pulse, but comprised a large number of individual spikes generated throughout the pulse. The frequency of spiking was typically in the order of 10–6 s for the pulses used in the experiment and consequently the power rating fluctuated over an appreciable range of values with a frequency in the MHz range. The results for power intensity were difficult to interpret unless these fluctuations were averaged out over several spikes. The averaging was achieved by recording a voltage proportional to the pulse energy incident on the photodetector to that instant, instead of a voltage proportional to the power [8]. The fluctuations produced only a high frequency ripple in the former voltage signal and the average rate of change with time of the signal was discernable, giving a quantity proportional to the effective instantaneous power intensity. The detail analysis of the temporal variation of power output from the laser and the power intensity distribution in the focal region were presented in the previous study [8]. In addition, a care was taken to minimize the possible source of variation during the experimentation. In order to reduce the intensity of the beam sampled, an attenuator was used, i.e. the power rating of the reference beam was reduced to the order of 10^{-3} W. This was necessary to (i) protect the photodetector active area from the reference beam, and (ii) minimize the effect of the reference beam on the interaction mechanism taking place between the incident beam and the vapor plume, i.e., if the reference beam intensity is high it effects the plume characteristics such as the plume energy is increased due to partially attenuation of the reference beam. The reference beam was focused by a 51 mm nominal focal length lens positioned so that the focal plane passed as close to the axis of the incident beam as possible. The reference beam after passing the plume was collimated by a second lens and directed on to a spectrometer-light detector unit, which provided an output signal proportional to the incident power deposited on a photodetector. The grating spectrometer was used to eliminate all the wavelengths of emitted light from the plume, except 1.06 μm radiation, from reaching the photodetector unit, i.e. the spectrometer provided that 1.06 μm radiation was received by the photodetector unit. A mechanical iris was set into the screening to allow the reference beam to be transmitted onto the photodetector. The iris aperture was adjusted to a diameter slightly exceeding the reference beam. Each measurement result, for a particular set of conditions, was recorded using a storage oscilloscope as a trace showing the fluctuations in the transmitted reference beam power during the incident laser beam workpiece interaction. The reduction in transmission of reference beam power due to partially blocking, defocusing, attenuation and scattering by the plume was accordingly monitored. To change the position of the reference beam relative to the workpiece surface, the reference beam was moved using a positioning micrometer. The position of the incident beam focusing lens relative to the target surface was also monitored using a gauge sensor as indicated in the previous study [8]. The initial location of the reference beam was obtained by noting the position of the workpiece when the beam could be seen to impinge on the workpiece. To allow results to be obtained for positions close to the target surface, narrow targets nominally 3–4 mm wide were used. This method prevents blocking of the reference beam by the workpiece.

The light detector was also used to indicate some features of the optical emission from the plume. In this case, the grating spectrometer was removed. The detector unit and screening were arranged to maximize the thermal emission seen by the photodetector, i.e. the photodetector was exposed to a large portion of the plume, giving output for spatially integrated emission, by opening the iris aperture to its maximum diameter. However, closing the iris aperture to a small diameter reduced the size of the plume exposed to the photodetector to give some degree of spatial resolution. The streak photography was carried out using a drum camera rotating a 1.5 m length of film at approximately 3000 rpm. This ensured that detail of the initial ejection from the laser produced cavity was observed by a streak persisting longer than the period corresponding to complete revolution. The delay between triggering and the occurrence of laser output allowed the shutter to open completely before any exposure of the film. After the first trial of the experiment, photographs were examined and the magnification was adjusted so that the space between the lens cover slide and the workpiece filled the film. This allowed observation of qualitative detail during the cavity formation and that gave information about the major phenomena involved.

4.4 Findings and Discussions

The results presented are associated with the above surface characteristics of the plasma and vapor plume in line with the previous studies [9, 10]. A streak photography capturing the evaporation process is shown in Fig. 4.5. It can be observed that several separate phases of the process occur. Initially, vapor leaves the surface and expands and it decelerates while it expands. In this case, the vapor front appears in highly luminous flares. In some cases, this appears as a series of small spikes, which initially appear to be about 10 μs apart, but after about 300 μs from the start of the pulse they fall of considerably in frequency. It is possible between the flares that plasma expands rapidly at which time the power intensity is still high. There also appear to be two types of liquid ejection in general, although the distinction between the two is only clear for nickel. The first kind of ejection starts early and covers the radially ejected slow particles. The second type of ejection starts a little later (100 μs) and comprises the fast particles ejected (these forming the steepest angles on the photographs). These particles seem to be very fine and are ejected almost continuously, as they give an almost uniform exposure of the film. A plot of the time of ejection of the last particle for each case is shown in Fig. 4.5. There is little correlation which exists with the laser energy. It should be noted that in all cases particles were ejected just after penetration of the workpiece suggesting that there was still some amount of liquid existing at the hole bottom at that time. Figure 4.6 shows the very tail of the streak photograph corresponding to nickel: in this case, liquid expulsion is evident. This appears as radially developed straight lines and small spots in the figure.

Fig. 4.5 A streak photograph

125 μs

Fig. 4.6 Streak photograph
showing the liquid ejection
and the bright spots are liquid
globules

Fig. 4.7 Temporal variation
of vapor front height

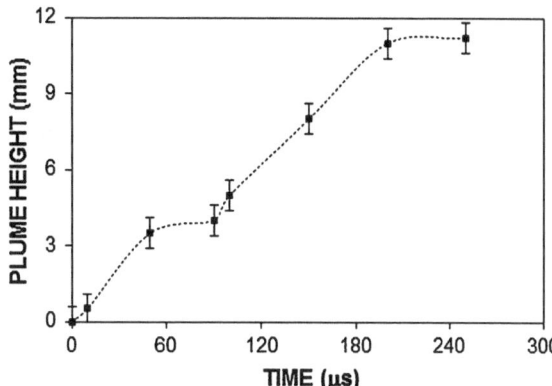

This may occur because of the ejection of small quantities of either vapor or liquid phases. The ejection mechanism is possibly corresponding to intense and rapidly expanding evaporating front, which is produced when the cavity is shallow, at which the time the power intensity is high.

Figure 4.7 shows the vapor front height (plume height) where the probe was positioned at 0.3 mm away from the central axis of the heated spot. The results are obtained from the Langmuir probe. The plume height increases with progressing heating period. It should be noted that the measurements are terminated after the time scale of 250 μs due to experimental difficulties. The plume height reaches 11.2 mm at the heating period of 250 μs. Moreover, the smooth variation of plume height with time is disturbed at about 100 μs of the heating period. This indicates that the amount of vapor ejected is reduced due to less energy reaching the workpiece surface. In this case, evaporation rates reduces, which in turn lowers the plume height.

Figure 4.8 shows P_l/P_v (P_l is the pressure in the liquid phase and P_v is the pressure in the vapor phase, $\frac{P_l}{P_v} = 1 + \frac{\rho_v V_u}{\rho_u} \sqrt{\frac{\gamma P_u}{\rho_u}}$ [6]) with surface temperature. It is evident from the figure that liquid pressure is of the order of 1.5 times the standard atmospheric pressure and the variation of Pl/Pu with surface temperature at the interface is negligible. This is due to the fact that as the surface temperature rises, both liquid and vapor phase temperatures rise at the interface: in this case, the liquid density does not change and vapor pressure is almost equal to atmospheric pressure, i.e. vapor expands into an atmospheric environment. Once the surface

Fig. 4.8 P_l/P_v with surface temperature for nickel and copper

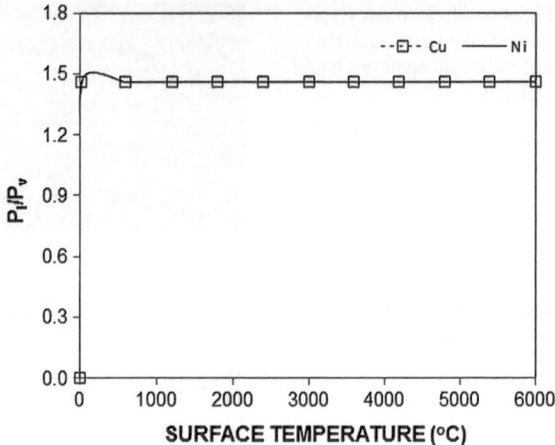

temperature builds up (i.e. 3000 °C and above), the surface evaporation velocity remains almost constant. Since the bubble formation occurs just below the surface, this increases the pressure in the liquid when it bursts, which in turn results in increased velocity of the particles ejected, i.e. causing fast particle ejection.

4.4.1 Plasma Absorption

The transmittance characteristic of the plume is probed at various distances from the target surface with the reference beam. The transmitted instantaneous reference beam power passing through the plume is monitored as a function of time. The tracers are reproduced in Figs. 4.9 and 4.10, showing the percentage transmission of the power as a function of time. However, the beam transmittance traces appear to be complex making the comparison difficult, but some trace characteristics are sufficiently distinct to provide a basis for assessment, although it is necessary to enclose the original results. Three characteristic regimes may be identified, and these are indicated in Fig. 4.11. Regime A is characteristic of the early part of the beam transmission period. The early part of the traces has a steep front indicating a rapid rise in transmittance as the leading edge of the plume reaches the reference beam. There follows a period of trace in which appear the maxima and minima of several spikes. The minima are associated with successive bursts of plume, which retain their identity during expansion. The maxima in the transmittance indicate the trailing edges of some vapor bursts are overtaken by the following bursts. In regime B, diffusion effects start to appear as the result of elapsed time after ejection or a slowed ejection of the vapor and liquid substrate. The trace typically exhibits large differences between peaks and troughs in percentage transmission, each of which envelop more than one of the spikes prominent in regime A. Regime C is characterized by a smoothed trace which displays ripples

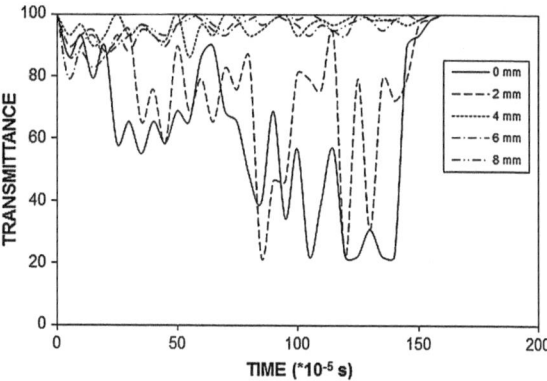

Fig. 4.9 Temporal variation of transmittance of reference beam for nickel (15 J and beam waist position 2.5 mm)

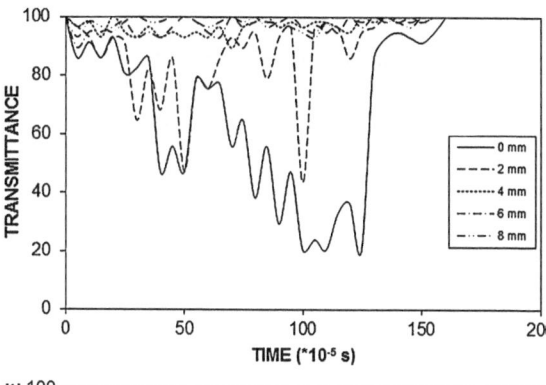

Fig. 4.10 Temporal variation of transmittance of reference beam for copper (15 J and beam waist position 2.5 mm)

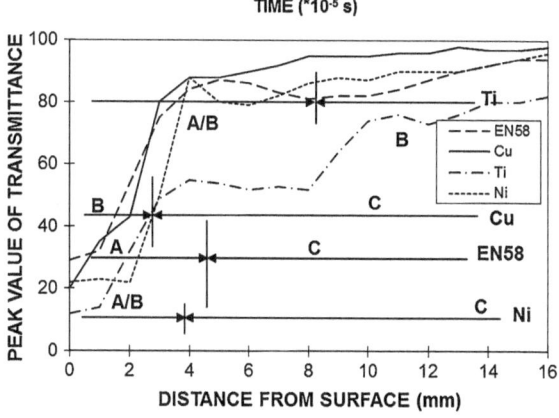

Fig. 4.11 Peak values of transmittance with the reference beam position from the surface for different materials (15 J and b.w.p. 2.5 mm). *A* Strongly defined spikes, *B* Transition, *C* Smoothed/Rippled

in contrast to large peaks and troughs. The transmittance of the reference beam is typically a large percent of the maximum possible.

Often different parts of the same transmittance trace may be associated with different respective regimes. For positions near the workpiece surface the first part of the interaction produces fine scale fluctuations in the reference beam transmittance as a characteristic of regime A, but towards the end of the trace

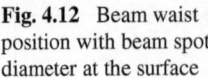

Fig. 4.12 Beam waist
position with beam spot
diameter at the surface

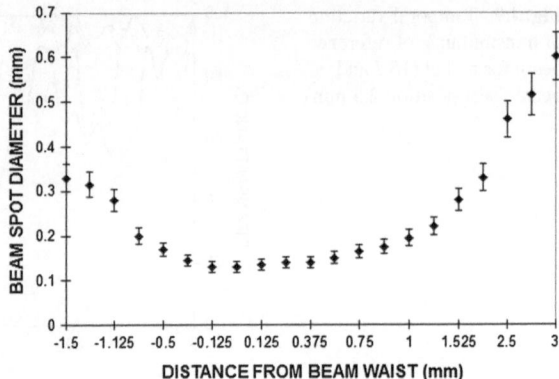

the transmittance is better described as a characteristic of regime B. The transition indicates changes in the plume at a fixed position which are attributed to the combined effects of changes in the cavity geometry and power dissipation at the surface, and time dependent processes such as heat diffusion into the workpiece material. With increasing distance from the surface, the recorded transmittance traces become smoother and a transmission peak in the initial part of the trace becomes prominent. The latter feature is at least partly due to the leading edge of the plume spreading radially and presenting a longer path length through the plume for the reference beam.

4.4.1.1 Effect of Changing Laser Beam Waist Position

The beam waist position was varied from 2.54 to 0.76 mm thus giving an increase in power intensity from 10^{11} W/m^2 to approximately $6* 10^{11}$ W/m^2 at the workpiece surface and a reduction in the beam diameter from 0.48 to 0.15 mm at the surface (a power intensity distribution with beam waist position is shown in Fig. 4.12 as obtained from the previous study [8]). A series of transmittance test results were obtained for stainless steel workpieces which are shown in Fig. 4.13.

Fig. 4.13 Temporal variation
of transmittance of reference
beam for steel (EN58) at two
beam waist positions (15 J,
beam waist positions are 0.8
and 2.5 mm)

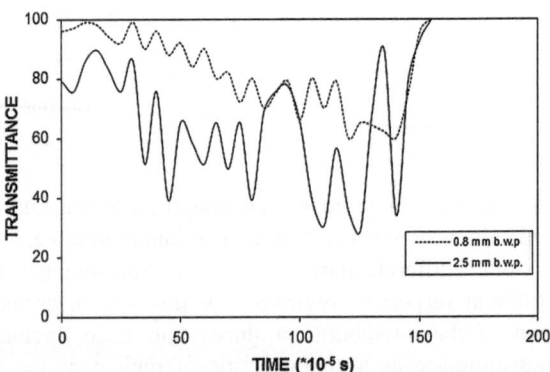

The vapor plume yields considerably low reference beam transmission only when the reference beam is positioned close to the surface (1 mm or closer). The most frequent fluctuations in transmittance produce peaks at about 20 μs intervals. Although focusing the laser beam to the workpiece surface gives a high power intensity at the surface, the plume generated by the interaction produce more transmission of the reference beam. This indicates that spot size and cavity geometry are at least as important as power intensity in determining the amount and state of the ejected substrate.

4.4.1.2 Effect of Changing Pulse Energy

The laser output pulse energy was varied during a series of laser pulses for which the other variables were kept constant. The pulse length, workpiece material, and beam waist position are 1.4 ms, titanium, and 2.5 mm above the workpiece surface respectively. The plume was probed at a position 3 mm above the workpiece surface by the reference beam. The results are shown in Fig. 4.14. The range of pulse energies investigated is 7–20 J. The effective absorption produced by vapor plume is observed to increase with pulse energy up to 13 J and remains constant in the range of 13–20 J.

The form of the absorption traces is unchanged in the 13–20 J pulse energy range. This shows rapid increase in absorption over about 40 μs when the leading edge of the plume reaches the position of the reference beam. For the first 400 μs of the transmission period, the plume produces fine scale fluctuations, with a periodic time of about 20 μs, about a low percentage mean transmittance of the reference beam. After this period, the trace was broken up into periods of low and high transmittance each lasting times of the order of 100 μs, which is characteristic of regime B (Fig. 4.15). Figure 4.15 shows the peak transmittance with laser output energy for titanium workpiece. As the pulse energy is reduced in increments of from 13 to 7 J, the average level of transmittance throughout the period of the interaction increases, particularly during the first half of the period. Towards the end of the interaction, low transmittance peaks may still be resolved showing that

Fig. 4.14 Effect of laser output energy on temporal variation of absorption of reference beam at 3 mm above from the surface for titanium (beam waist position 2.5 mm)

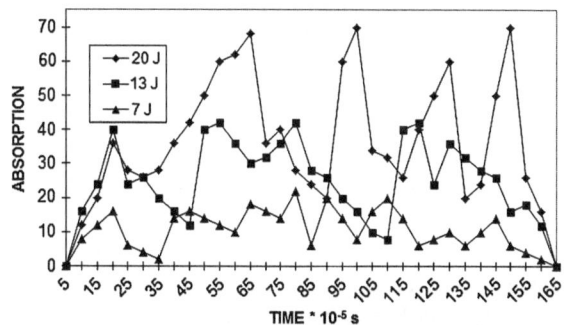

Fig. 4.15 Variation of peak
transmittance with laser
output energy for titanium
for reference beam position
of 3 mm above the surface
(beam waist position 2.5 mm)

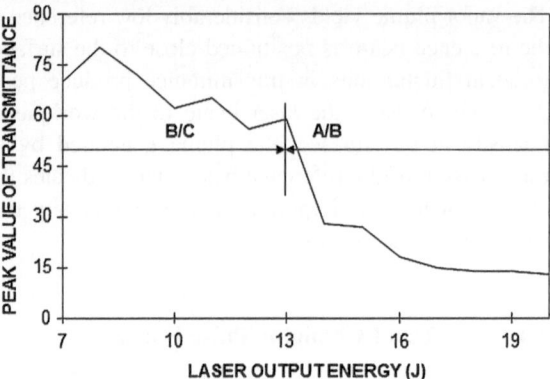

although the amount of material ejected from the target is reduced the plume is
maintained in a partially blocking state. The transmittance traces display fine scale
fluctuations for the period corresponding to the first half of the plume reference
beam interaction throughout the range of the pulse energies employed. This indicates that plume emission occurs in bursts throughout the range of pulse energies
used, including the lower limit of 7 J, which corresponds to a power intensity at
the workpiece surface of the order of 10^{10} W/m^2.

4.4.1.3 Source of Partial Blocking and Scattering

Although material ejected from the workpiece produces partial blocking and scattering of the reference beam, in the first instance it is unreasonable to attribute the
effect either to the presence of the high density vapor or liquid globules. However,
an assessment of the results shows that it is the high density vapor, which produces the partial blocking and scattering. Two indications are visible in this case.
Firstly, fine scale fluctuations in transmittance are observed for the reference beam
positions close to the surface have a frequency in the range of 50–100 peaks/ms
and this corresponds to the frequency of vapor bursts from the target surface.
Secondly, if the low transmittance spikes are due to single particles, the spikes
should become less frequent, whilst retaining a similar spike shape, with increasing reference beam position from the surface due to a decrease in particle number
density. For the same reason, if spikes are caused by the bursts of particles, the
spikes should tend to last for shorter times rather than larger times as observed
with increasing distance of the reference beam from the workpiece surface since
only the vapor plume is appreciably decelerated by the surrounding gas.

The four processes, which may contribute significantly to the reference beam
transmittance, are partial blocking, absorption, scattering and reflection. However,
no significant reference beam reflection could be detected along or close to the
axis of the reference beam during interaction. Projecting the transmitted reference
beam over an 8 mm path length onto a screen and photographing the beam spot

during interactions produces no evidence of incident beam movement or diffusion occurred, such as photograph of streaks or an increased spot size. Moreover, the power intensity used in the present experiment produces the electron number density in the order of 10^{17} cm^{-3} [5], which in turn results in the absorption coefficient in the order of 0.4 m^{-1}. Consequently, plume only absorbs the small fraction of the incident beam. These failures to detect reflection or scattering and large absorption depth, although non-conclusive, suggest partial blocking due to high density evaporating front is the dominant process.

4.4.1.4 Emission from Titanium Plume Detected by the Light Detector Unit

The photodetector was screened from the titanium plume except for a small aperture of 0.6 mm diameter positioned relative to the detector unit and the plume. The arrangement exposed the active surface of the photodetector to plume emission from a region 2.2 mm in diameter in the plane of the incident laser beam axis, providing some spatial resolution of the source of the emission from the plume to be seen by the detector. The results are presented in Fig. 4.16. Strong radiation emission occurs over a shorter period than that during which low reference beam transmittance occurs. The difference is greatest near the surface where low transmittance occurs over a period 0.8–1.2 ms compared to a period of strong emission lasting only 0.4 ms corresponding to the first half of the transmission period. The high radiation emission and low transmittance may indicate the existing of high density state of the plume.

For positions of the photodetector/aperture axis equal to or less than 4 mm above the target surface the emitted power rises steeply to its peak value at 100–150 µs from the start of emission followed by a larger period of decline interrupted by a less number of smaller peaks. Fine-scale fluctuations in radiated power were monitored for positions near the surface, occurring at 18–22 µs intervals, which is the same period associated with fine scale fluctuations in transmittance. The peak power radiated from the plume is observed to decrease with distance

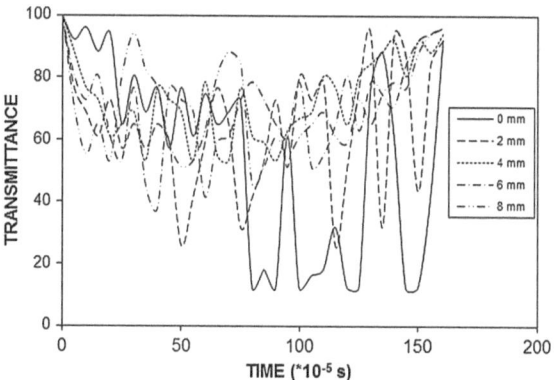

Fig. 4.16 Temporal variation of transmittance of reference beam for titanium (15 J and beam waist position 2.5 mm)

from the workpiece surface of the photodetector/aperture axis. The peak value decreased by 75 % over the range of positions up to 10 mm away from the surface. For positions up to 5 mm from the surface the emission produces several sharp spikes in each trace. These spikes are smoothed out for positions equal to or greater than 10 mm from the surface and the period of emission increases indicating the occurrence of diffusion and deceleration of the plume.

4.4.1.5 Emission Characteristics for Different Workpiece Materials

By increasing the aperture diameter to approximately 2 mm, the plume size exposed to the photodetector substantially increases allowing the radiation emission from a range of target materials, including some that radiate only weakly, to be monitored under the same conditions by changing the target material between laser pulses.

The active area of the photodetector was exposed to radiation from plume in a region 4.3 mm in diameter in the plane of the incident laser beam axis. The photodetector/aperture axis was aligned to pass through the plume 5 mm above the target surface. Four workpiece materials were employed, namely titanium, stainless steel, nickel and copper. The test results are shown in Fig. 4.17. Each material produced distinguished emission characteristics of temporal evolution and intensity. The differences between the radiated power intensity of respective target materials may be expected not least because of observed differences in workpiece mass removal rates and liquid ejection temperatures described in the early studies [10].

Differences between the time resolved form of the radiated power trace for respective target materials may reflect differences in the evolution of the damage crater. Since the emission characteristics are associated with periods longer than those directly related to machining process, the characteristics are more likely indications of changes in shape and size of the damage crater and changes in machining rates averaged over 10–100 μs. Titanium has a uniform emission characteristic, which suggests a uniform rate of machining. Similarly, copper emissions build up to its maximum intensity over 400 μs whereas stainless steel

Fig. 4.17 Temporal variation of transmittance of reference beam for different materials (15 J, beam waist position 2.5 mm and 0 mm)

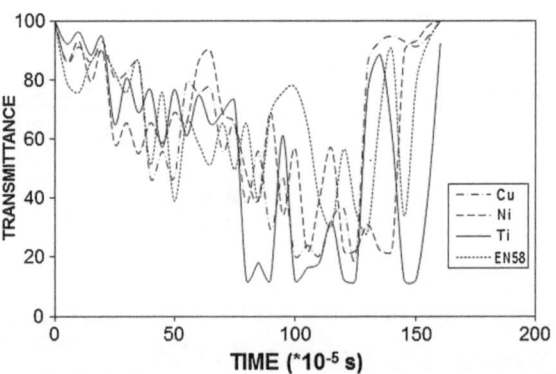

emission builds up over a 100 μs period suggesting the latter workpiece material allows the machining rate to reach a maximum for the interaction more rapidly than the former material.

The streak photographs (Fig. 4.5) reveals that several separate phases of the ejection process may be seen. Initially vapor is ejected from the hole and decelerates rapidly as it expands. In many cases, the vapor front comprises the envelope of several dense plumes of vapor, that is, additional vapor is not produced continuously but in pulses. Series of small spikes appear on the workpiece surface about 10 μs apart, and their frequency falls off considerably after about 300 μs from the start of the laser heating pulse. These spikes do not seem to correspond to ejection of large quantities of vapor and/or liquid; they are regions of intense rapidly expanding vapor fronts, which are produced only when the cavity is shallow, at which time the power intensity is still high. It should be noted that the intensity on the workpiece surface reduces as the surface recedes, since it is moving further out of focus. The slowest traces on the film (those forming the shallowest angle) can be produced back to the point where vapor ejection commences, showing that these particles are the first to be expelled. In fact, some traces appear to start before the vapor in some cases. This may occur because; (i) the particles decelerate and so give rise to curved traces where the trace is hidden by vapor, thus only that part some distance from the point of initial vapor ejection may be produced back and this may make the particle appear to start too early, and (ii) some particle traces originate in a dense vapor and may be supposed to come from an explosion of another particle. Some of these traces actually move towards the workpiece surface from a vapor plume. Some curved traces are obtained in which a particle changes direction due to evaporation from the surface nearest the incoming beam.

The second type of ejection starts a little later (~100 μs) and comprises the fastest particles ejected (those forming the steepest angles on the photographs). These seem to be very fine and are ejected almost continuously as they give an almost uniform exposure of the film. During the ejection of this type of particle, a faster particle of the first type starts to be expelled. They are apparently much larger than the fastest variety as the tracks they leave are up to 1 mm wide, and are of greatly varying velocities. A plot of the time of ejection of the last particle for each case is shown in Fig. 4.18. There appears to be little correlation with the laser

Fig. 4.18 Time for last particle ejection from the cavity with laser output energy (beam waist position 2.5 mm)

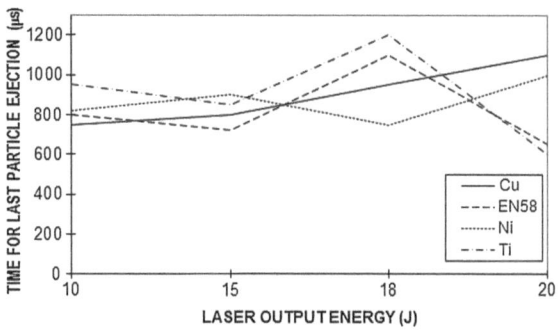

energy level, although the focus effect, which is included, may obscure this. The fact that in all but five cases, particles were ejected after penetration of the workpiece suggests that there was still a considerable amount of liquid at the bottom of the cavity at that time.

It was postulated that the earliest particles of the first type of liquid ejection are caused by radial flow [10]. The initial ejection due to radial flow should produce fairly slow particles since early in the pulse the pressure gradients in the cavity are small. Later, the increasing pressure gradients give rise to faster particles. When explosions occur, high velocities may be achieved provided the ejection particles are small. As the pulse progresses, larger particles may be exploded giving wider slower streaks. It seems reasonable to assume that the last particles ejected, which give such streaks, are produced by explosions, since radial flow plays a decreasing role later in the pulse. Each of the four materials examined was found to give rise to a characteristic streak photograph; the major variation was in quantity of each type of ejection.

References

1. Yilbas BS, Davies R, Yilbas Z, Gorur A, Acar I (1992) Surface line and plug flow models governing laser produced vapor from metallic surfaces. Pramana J Phys 38(2):195–209
2. Yilbas BS, Mansour SB (2006) Laser evaporative heating of surface: simulation of flow field in the laser produced cavity. J Phys D: Appl Phys 39(17):3863–3875
3. Yilbas BS, Davies R, Yilbas Z, Begh F (1990) Analysis of absorption mechanism during laser metal interaction. Pramana J Phys 34(6):473–489
4. Yilbas BS (1987) Study of affecting parameters in laser hole drilling of sheet metals. Trans ASME J Eng. Mater Technol 109:282–287
5. Yilbas BS, Yilbas Z (1987) Plasma transients during laser drilling in subatmospheric pressure atmospheres of air. Opt Lasers Eng 7:1–13
6. Yilbas BS, Davies R, Yilbas Z, Begh F (1990) Study into the measurement and prediction of penetration time during CO_2 laser cutting process. Proc Inst Mech. Eng Part B: J. Eng Manuf 204:105–113
7. Yilbas BS (1986) The absorption of incident beams during laser drilling of metals. Opt Laser Technol 18:27–32
8. Yilbas BS, Yilbas Z (1996) An optical method to measure the pulsed laser output power intensity distribution in the focal region, measurement. J Int Meas Confed 17(3):161–172
9. Yilbas BS, Sami M (1997) Liquid ejection and possible nucleate boiling mechanisms in relation to the laser drilling process. J Phys D: Appl Phys 30:1996–2005
10. Yilbas BS (1995) Study of liquid and vapor ejection process during laser drilling of metals. J Laser Appl 17(3):147–152

Chapter 5
Concluding Remarks

Laser drilling involves with multiphysics phenomena, which require careful examination of each physical process for improved end product quality. The laser workpiece interaction is governed by solid heating, phase change and plasma formation at the surface. Since the physical processes are complicated, the closed form solutions incorporating the analytical methods do not provide the accurate results. Therefore, the numerical investigation of the drilling process becomes unavoidable. However, the processes can be simplified to focus around the temperature, thermal stress, and plasma absorption effects. Although the coupling among these effects is significant, the influence of each effect on the process can be possibly examined through the numerical simulations and experimentations of the actual processes. On the other hand, the end product quality can be assessed through the close examination of hole geometry. Since affecting parameters are many, the statistical approach incorporating the factorial design becomes necessary prior to drilling experiment. Therefore, in this book, three main focus areas of research are presented, namely numerical modelling, factorial analysis and the plasma plume and its affects on the drilling. The conclusions derived from each focus area are presented seperately below under the appropriate sub-headings.

5.1 Model Studies

The diffusional energy transport governs the temperature field for the temperatures less than the evaporation temperature of the irradiated material. Once the evaporation initiates, interface temperature remains at evaporation temperature and temperature across the interface in the solid as well as in the vapor sides decays sharply due to laser moving heat source located at vapor–solid interface. The surface recession due to evaporation is high in the early heating period. As the heating progresses further, recession velocity of solid surface becomes almost constant, i.e. does not change with progressing time. Consequently, transient affect of evaporation becomes apparent in the early duration of evaporation, and in the later stages it remains almost steady despite the process is transient. Since the vapor emanating from the laser produced cavity expands into

B. S. Yilbas, *Laser Drilling*, SpringerBriefs in Manufacturing and Surface Engineering, DOI: 10.1007/978-3-642-34982-9_5, © The Author(s) 2013

its ambient, low density variation between vapor phase and its ambient slow the vapor expansion. This results in complex flow structure within the cavity. Consequently, radial expansion of vapour front enhances the pressure increase in the region of cavity edges. Since the rate of evaporation varies with time, pressure build up in cavity edges also varies with time. The cavity extension along the symmetry axis is larger than that corresponding to the cavity edge. This is due to absorption of laser power intensity, which is high along the symmetry axis. The size of the mushy zones in the region close to the cavity edge is smaller than that of symmetry axis. The size of solid–liquid mushy zone is smaller than the size of liquid–vapor mushy zone.

The flow field, due to the assisting gas, at the hole inlet is influenced by the streamline curvature of the impinging jet. In this case, radial momentum enables fluid in the assisting gas to flow towards the hole wall region. This is more pronounced in the region close to the edge of the flow. The flow field in the entrance region of the hole is influenced by the blockage effect of the hole. The influence of the hole wall velocity on the Nusselt number is not significant, except as a small effect is observed at the hole inlet and the exit regions. In this case, the Nusselt number increases slightly with increasing the hole wall velocity at the entry region of the hole while it is reduced by increasing the hole velocity at the exit region of the hole. Increasing the jet velocity increases the Nusselt number along the hole wall. The skin friction attains high values at the hole inlet and the exit. Increasing jet velocity enhances the skin friction at the hole inlet region while it is opposite at the hole exit region.

von Mises stress shows a wavy behavior in the radial direction. This is because of the level of stress components in the radial direction, i.e. tangential stress component changes from compressive to tensile in the region close to the hole wall while the trend of the axial and radial stress component remains the same along the radial direction. The von Mises stress distribution in the axial direction is similar to the axial stress distribution. The von Mises stress rises rapidly in the early heating period and as the heating period progresses the rise of equivalent stress becomes steady with time, i.e. it becomes similar to the temporal variation of the temperature, provided that they have different magnitude of gradients. The von Mises stress attained relatively low values in the vicinity of the top and bottom surfaces of the workpiece because of the free expansion of this surface in comparison to the hole mid-section. However, the von Mises stress was significantly higher in the hole wall surface vicinity, which became more pronounced at the y-axis location of the mid-hole height. This attainment of high values of von Mises stress is attributed to the strain developed due to expansion and compression of the substrate material in this region. However, as the distance along the x-axis increased from the hole wall surface vicinity, the von Mises stress reduced sharply. As the distance increased further, this decrease became more gradual. It is evident from the hole micrographs that some small surface debris occurred around the hole circumference on the top and bottom surfaces of the workpiece. This is attributed to the drag forces developed in the hole vicinity during the drilling process. Parallel sided holes were drilled and gave rise to the formation of some small hole inlet and exit cone angles, caused by material expulsion during the drilling process.

The minimum specific energy requirements for cutting reduces with increasing hole diameter. This is because of the increase in the volume of the material removed at large hole diameters during the cutting process. The first and second law efficiencies improve significantly at low levels of laser power and high laser cutting speeds. The second law efficiency remains lower than the first law efficiency. This is because of the entropy generation through heat transfer during the cutting section. The trends of the second law efficiency follows almost the first law efficiency, i.e., increasing laser power while lowering cutting speed lowers the second law efficiency. The experimental findings reveal that increasing laser power while lowering the cutting speed results in sideways burning and cutting irregularities around the hole circumference. The sideways burning is associated with the high temperature combustion process. Consequently, improving the first and second law efficiencies enhances the end product quality of the laser cut holes.

5.2 Assessment of Hole Quality

The effects of thickness, focus setting, and ambient pressure are identified as the most significant parameters. The effects of thickness and focus setting are most significant for overall quality of the hole drilled. The main effect of pressure is found to be most significant on taper, mean hole diameter while it is significant for resolidified material, barreling, and overall quality. The first order interactions of pressure–focus setting, and pressure–thickness indicate that the coupling effects of these parameters are significant for geometric features of holes drilled. This is especially true for resolidified material, taper, inlet cone, mean hole diameter, and overall quality. Focus setting effects significantly all the geometric features of the hole geometry, except resolidified material and surface debris. In addition, its first order interactions with pressure, thickness, and energy have significant effect on the hole geometry. The main effect of energy is found to be significant for taper, barreling, and overall quality. However, its first order interaction with pressure has no significant effect on the hole geometry. The main effects of thickness and its interactions have significant effect on the features of the hole geometry. This indicates that the thickness of the workpiece is one of the most important parameters influencing the shape of the hole geometry. From the qualitative analysis, the effect of each parameter on the hole geometry varies almost linearly, except thickness and focus setting which vary quadratic or cubic forms. This indicates that small variation of pressure and energy result in small changes in hole geometry, but this change amplifies once the thickness and focus setting vary.

5.3 Evaporating Surface and Plasma Plume Absorption and Transmissitivity

Evaporating front velocity is influenced by the mass removal rate. In this case, energy reaching the workpiece surface is partially blocked by the vapor front, which in turn reduces the evaporating front velocity. In the early heating period, the velocity profile appears to be a triangle shape and as the heating progresses it becomes almost fully developed turbulent velocity profile. This may occur because of the expansion of the vapor front to its surrounding.

The plume height measured is as high as 11.2 mm at 250 μs of the heating duration during the drilling. The plume height is also influenced by the mass removal rate. Close to the workpiece surface, the plume generated during interaction produces large scale fluctuations in the reference beam transmittance for all the workpiece materials employed. Consequently, in general, the plume reduces the incident power reaching the workpiece surface appreciably through partially blocking the incident beam. The observed reference beam transmittance is attributed to dense vapor effects as compared to blockage by liquid particles ejected from the target. The reflection, deflection and absorption of the incident beam by the plume are unlikely. The variation of transmittance with time and the position of the reference beam relative to the surface indicate variations in the dense vapor state and rate of expansion occur. The effects of changing workpiece material, incident laser beam waist position and laser pulse energy were investigated in turn. The results conclude that each of these variables influences the transmittance properties of the plume. The thermal radiation emission from titanium plume was monitored for a range of positions above the workpiece surface. The results indicate that the radiated power decreases more rapidly than the transmittance capability of the plume both with time and distance from the workpiece surface. Early in the interaction, for positions close to the workpiece surface, the radiated power fluctuates with time in a similar way to the reference beam transmittance. The fluctuation is associated with a non-steady rate of dense vapor and liquid ejection from the workpiece. By exposing a sufficiently large volume of the vapor to the photodetector used to monitor the emission, a comparison of results was possible for different materials including those radiating at low energy. Under similar conditions, each material vapor radiates power as a function of time, which is peculiar to that material. Initially very slow particles are ejected at the same time that dense vapor ejection commences. These slow particles are probably the result of liquid flowing radially out of the cavity due to radially dependent recoil pressure from evaporating material at the surface. As the laser heating pulse progresses, the pressure gradient, and hence the velocity of the liquid, increases. Consequently, a situation arises where nucleation within the liquid zone produces explosions leading at first to small, fast particles and later some larger, slower particles. After the expulsion of the fastest particles, there then follows, in most cases, a reduction in the dense vapors ejection, and then a sudden resumption. The major contributions to this second burst of vapor appear to originate, not from within the material, but above the surface, suggesting that evaporating liquid particles are responsible. This implies that the dense vapor leaves the workpiece after the initial burst, with some of the material being liquid at the time of ejection.